Scientific Sheep Farming

SCIENTIFIC SHEEP FARMING

EDITORS :

Mir Nadeem Hassan

Iqra Hussain Shah

Showkat Ahmad Shah

Ajaz Ahmad Ganie

Zahoor Ahmad Wani

NARENDRA PUBLISHING HOUSE

New Delhi-110 085

First Edition 2023

ISBN : 978-93-56512-43-6

Published by:

NARENDRA PUBLISHING HOUSE

Publisher and Distributor

C-21, Varun Apartments, Sector-9
Rohini, Delhi- 110085 (INDIA)
Ph: +91-11-45025794, +919717874875
Email: info@nphindia.com, nphindia@gmail.com
Website: www.nphindia.com

Printed in India

Composed & Designed by Palak Graphic Designer

Contents

Preface ..... *vii*

About the Editors ..... *ix*

1. Normal Body Parameters, Common Terminology Used & Dentition in Sheep 1
   Andleeb Rafiq, Showkat Ahmad Shah
2. Basic Farm Operations 5
   Muzamil Abdullah
3. Shelter Management for Sheep 9
   Raoof A. Patoo, Irfan A. Baba, Muzamil Abdullah, Ashiq manzoor, Majid S Kawoosa
4. Pregnancy Diagnosis, Management of Pregnant Ewes and New Born Lambs 15
   Asloob Ahmad Malik
5. Feeding Management of Sheep 19
   Ajaz Ahmad Ganie, Mir Nadeem Hassan, Manzoor Ahmad Yatoo, Iqra Hussain Shah
6. Factors Affecting Increased Fertility and Overall Performance of Sheep in Light of Genetics 25
   Shabir Mir, Aijaz Sydain, Nusrat Nabi, Majid Shafi, Saba Bukhari
7. Care and Management of Different Age Groups of Sheep 29
   Iqra Hussain Shah, Mir Nadeem Hassan, Muzamil Abdullah, Showkat Ahmad Shah, Ajaz Ahmad Ganie
8. Important Sheep Diseases and their Management 35
   Showkat Ahmad Shah, Majid Shafi, Syed Ashaq Hussain, Henna Wani, Mir Nadeem Hassan
9. Footrot in Sheep - its Prevention and Control 43
   Shaheen Farooq, Mir Nadeem Hassan, Iqra Hussain Shah

10. Importance of Deworming in Sheep & Goat 49
Zahoor A. Wani, Idrees M. Allaie, Kamal H. Bulbul, Shahana R. Tramboo, Insha Akbar, Showkat Ahmad Shah

11. Common Parasitic Diseases of Sheep and their Treatment and Control 55
Idrees M. Allaie, Zahoor A. Wani, Kamal H. Bulbul, Aiman Ashraf and Mufeeda Tariq

12. Commonly Occuring Plant Poisonings in Sheep and Coats in Kashimir 65
Syed Wasif Ahmad, Adil mehraj Khan, Showkat Ahmad Shah

13. Vaccination in Sheep 69
Mir Nadeem Hassan, Iqra Hussain Shah, Ajaz Ahmad Ganie, Shaheen Farooq

14. Abortions 73
Iqra Hussain Shah, Mir Nadeem Hassan, Asloob A. Malik, Showkat Ahmad Shah, Ajaz A. Ganie

15. Management Year Calendar for Small Ruminants 77
Mir Nadeem Hassan, Manzoor Ahmad Yatoo, Andleeb Rafiq, Ajaz A. Ganie, Showkat A. Shah, Iqra Hussain Shah

16. Methods of Drug Administration in Sheep 81
Syed Ashiq Hussain

17. Pasture Management 87
Dr Manzoor Ahmad Yatoo, Ajaz Ahmad Ganie, Mir Nadeem Hassan, Iqra Hussain Shah

18. Commonly Used Drugs in Sheep and Goats 91
Iqra Hussain Shah, Mir Nadeem Hassan, Syed Wasif, Syed Ashiq Hussain, Showkat Ahmad Shah

PREFACE

The book is an attempt to provide important information on almost all important aspects of scientific sheep rearing to breeders/entrepreneurs, rural youth associated with sheep rearing. The publication shall be useful for not only better disease control at field level but will also help improve profitability of livestock enterprise in terms of improvements in managemental practices and better nutrition. The book has been compiled from both published studies and field observations. Care has been taken to present the standard opinion on any given subject by citing the standard recommendations. This book provides breeders information on key inputs like feeding, dosing and vaccination schedules, updated sheep calendar, tips for record keeping and housing besides guidelines for most efficient production using latest scientific knowledge and inputs generated at field level.

This book "Scientific Sheep Farming" will be of great use to students, rural youth, and progressive farmers, extension workers, policy makers.

We are grateful to all those who have contributed to this publication in terms of articles and material and by putting their efforts in its compilation.

Mir Nadeem Hassan
Iqra Hussain Shah
Showkat Ahmad Shah
Ajaz Ahmad Ganie
Zahoor Ahmad Wani

ABOUT THE EDITORS

DR. MIR NADEEM HASSAN

Dr. Nadeem, working as Scientist (Animal Science) in SKUAST-Kashmir has 10 years of teaching, research and extension experience. Dr. Nadeem graduated from SKUAST-Jammu and obtained masters in Veterinary Microbiology and Immunology from GADVASU-Ludhiana and Ph. D from SKUAST-Kashmir. He is recipient of **S. S. Dhillon Gold medal** during M. V. Sc programme. He is also recipient of **Gold Medal** for securing first rank at university level during Ph. D. programme and **Merit-Cum-Means** Scholarship from Ministry of Minority Affairs Govt. of India. The author has also qualified **ICAR-Junior Research Fellowship (JRF)-All India 29th Rank**, National Eligibility test (NET) in paraclinical veterinary sciences. He has to his credit more than 40 papers published in National and International Scientific Journals. He has also handled **eleven externally funded projects** as PI/Co-PI.

DR. IQRA HUSSAIN SHAH

Dr. Iqra is working as Research Assistant in DST funded project on Foot & Mouth Disease. She graduated and obtained masters in and Ph. D in Veterinary Public Health from SKUAST-Kashmir. She has more than 10 research publication in national and international journals.

DR. SHOWKAT AHMAD SHAH

Dr. Showkat, works as Assistant Professor in the Division of Veterinary Pathology at SKUAST-Kashmir. He has 12 years of teaching, research and extension experience. Dr. Showkat graduated from SKUAST-Kashmir and obtained masters in Veterinary Pathology from GADVASU-Ludhiana and Ph. D from SKUAST-Kashmir.

DR. AJAZ AHMAD GANIE

Dr. Ajaz is Working as Scientist (Animal Science) at KVK/ETC Malangpora Pulwama SKUAST-Kashmir. He has been involved in extension, teaching and research for more than 10 years. Dr. Ajaz graduated from SKUAST-Kashmir and obtained masters in Animal Nutrition from JNKVV-Jabalpur and Ph. D from NDRI Karnal. He is a life member of Animal Association of India and Animal Nutrition Society of India.

DR. ZAHOOR AHMAD WANI

Dr. Zahoor, works as Assistant Professor in the Division of Veterinary Parasitology at SKUAST-Kashmir. He has 12 years of teaching, research and extension experience. Dr. Zahoor graduated from SKUAST-Jammu and obtained masters and Ph D in Veterinary Parasitology from SKUAST-Kashmir

1 Normal Body Parameters, Common Terminology Used & Dentition in Sheep

Andleeb Rafiq, Showkat Ahmad Shah

a) Normal Body Parameters

	Temperature	Respiration rate/minute	Pulse rate/minute
Average	102.5°F/ 39.1°C	19	75

b) *Common terminology in Sheep Rearing*

Ram	Male of breeding age (above 12 months)
Wether	Castrated male
Ram lamb	Immature male (8-12 months)
Ewe lamb	Immature female (12-18 months)
Lamb	Newborn till weaning
Flock	Group of sheep (3 or more)
Tupping	Act of mating male with female
Lambing	Act of parturition (giving birth)
Life span	10-12 years
Productive life span	2-8 years
Weaning age	3-5 months
Pasture carrying capacity	7 ewes & lambs/acre

Dentition

- Gives us an idea about the age of the animal
- Helps us in purchasing the sheep or goat
- Helps us to decide when to mate
- Helps us to decide when to cull the animal

Lambs at birth usually have no teeth or may present one pair of incisors. Within a week after birth, the milk teeth appear in the front lower jaw and by the time the lamb is two months old, all eight incisors erupt. The milk teeth are similar to the permanent teeth in form and placement but are smaller in size and fewer in number. These are subsequently replaced by permanent incisors, which appear in pairs, beginning with the two central teeth, followed by one on either side at intervals, until the eight milk teeth have been replaced. During the period the teeth are growing, sheep are referred to by the number of permanent incisors present, such as two-tooth, four-tooth, six-tooth, and eight-tooth or full mouth.

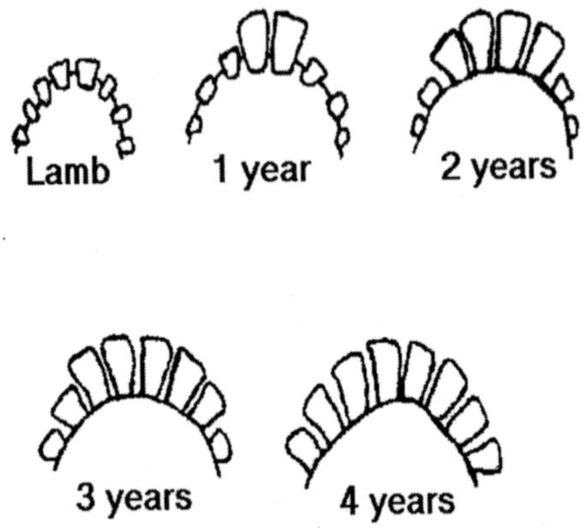

Diagram showing development of the permanent incisors and the variation in age at which they may erupt.

Age	Permanent teeth
1-1½ year	1st or middle pair of incisors
1½ - 2 years	2nd pair of incisors
2½ - 3 years	3rd pair of incisors
3½ - 4 years	4th pair of incisors
5 years	all permanent incisors close together
6 years	Incisors begin spreading apart
7-8 years	Some incisors broken
10-12 years	All incisors missing

Precise formula to remember dentition for both Sheep & Goat		
Estimated age range	**Sheep Goat**	
No. of permanent incisors		
0 pair	Less than 1 year	
1 pair	1-1½ years	1-2 years
2 pairs	1½-2years	2-3 years
3 pairs	2½-3years	3-4 years
4 pairs	More than three years	
Broken mouth	More than four years	

References

1. N.S.R. Sastry and C.K. Thomas. Livestock Production and Management. Fourth revised. Kalyani Publishers, Ludhiana, India.
2. Merck Veterinary Manual Online, 2011. Merk and Co.
3. Goat, Sheep and Pig, Production and management by Jagdish Prasad.

2 Basic Farm Operations

Muzamil Abdullah

The below mentioned farm operations are often missed by our progressive livestock breeders/ farmers/producers and this contributes to inefficiency, losses, diseases and other forms of mismanagement. The important farm operations to be followed at a sheep farm are as follows:

(a) Record maintaining

The most important in farm is maintaining recording of data of farm like performance of livestock, feeding, breeding, lambing/kidding, vaccination, dosing, death etc.

(b) Tagging

This is the most common method of identification of lambs. It is done to maintain records for ownership and selection purposes and is usually needed to keep track of treated animals until drug residues have cleared. Tags come from various materials and sizes. Plastic tags are better in many respects.

The electronic tags have been in vogue for quite some time.

(c) Crutching

Crutching consists of the removal of wool/hair from the areas around the vulva, udder, and the inside of the rear legs (usually done in sheep/goat). It can generally be done in any season because relatively little wool is removed. Crutching has the following benefits:

- Allows easy observation of the signs and progress of birth
- Reduces potential contamination of the lamb as it is born and of the ewe if assistance is needed.
- Makes it easier for lambs to find the teats to nurse.
- Reduces the potential contamination of the teats and udder.

- Makes it easier to observe whether the lamb has actually begun nursing.
- Reduces the potential for fly strike during warm seasons of the year.
- Increases the value of the fleece when sheep are shorn later.

(d) Hoof trimming

This involves removing overgrown hoof material. It keeps the hoof clean preventing diseases like abscesses. To avoid taking off too much hoof and causing bleeding or damage, the horn of the hoof should be cut back in several stages.

(e) Castration

Castration is done to achieve the following objectives:

- Prevent indiscriminate breeding, thereby exercising genetic control and regulating the lambing season.
- Prevents the breeding of young female flock mates that may become pregnant but are not in adequate physical condition for pregnancy and lactation.
- Prevent the development of aggressive behaviour in maturing males and the resulting injuries that frequently accompany this behaviour.
- If performed at a relatively young age, it enhances pelt removal and meat quality.

Note: The benefits of such procedures as castration, docking etc. outweigh the stress produced, provided a skilled person carries out the procedures carefully and efficiently.

(f) Tail docking

Tail docking is carried out in sheep/goat (in some animals) to achieve multiple objectives which include, among others

Disease Prevention: Obstructs development of fly strike which can be fatal sometimes. It allows better inspection of rear area of animal.

Aesthetic: Improving look

Hygienic: Reducing the possibility of soiling of the long tail with urine and faeces

The tail should be removed at the third tail joint. It means leaving the docked tail long enough that covers the vulva. Leaving the tail this length protects the vulva from the sun and flies. Docking any be performed at any day or time but preferably the weather should be sunny and age of lamb should be about 10 days. Castration, tail docking and ear marking are usually carried out together and are collectively known as lamb marking.

(g) Dipping

It is the dipping of small ruminants in medicated/ ectoparastidal water (in trenches/ tubs) doneto attain the following objectives:

- Control of ectoparasites and sheep scab
- Get clean wool, remove waste materials and dung from fleece
- Keep away sheep blow flies.

It is usually done once in a year before post winter shearing and before post autumn shearing. It is also done when the incidence of ectoparasites is high.

Methods include

1. Hand bath: it is used for small flock. Sheep are lifted one by one into a tank of galvanized iron (1.2 x1.0x0.5 m) and kept for 2 minutes. Sheep are then placed on drain board to drain off surplus dip back into tank.
2. Swim bath: it is used for large sized flock. Two to three sheep are dipped at a time.

Precautions include

1. Avoid dipping in advanced stage of pregnancy.
2. Always water the sheep before dipping to avoid drinking of dip solution.
3. Avoid dipping on rainy days so that dip may not be washed off.
4. Avoid dipping of sick animals, animals with wounds and young lambs.
5. Avoid dipping in rams in breeding season to guard from injury to penis.
6. Give drinking water to livestock before dipping to prevent drinking of this medicated water.

(h) Cleaning and biosecurity

Routine cleaning of livestock sheds and around the sheds and maintaining of proper draining is must for hygiene of farm. Avoiding mixing of outsiders both animals as well as human prevents spread of infection from outside farm.

References

1. A Textbook of Animal Husbandry by G.C. Banerjee. Eighth Edition.
2. Livestock Production Management . C K Thomas and N S R Shastry. Shastry Thomas.
3. Hand book of Animal Husbandry. V.K. Taneja, C. Chakravarty, C.S. Viswanath, Aruna T. Kumar. Third revised. Indian Council of Agricultural Research, New Delhi.

3 Shelter Management for Sheep

Raoof A Patoo, Irfan A Baba, Muzamil Abdullah, Ashiq Manzoor, Majid S Kawoosa

Normally sheep do not require elaborate housing facilities, but minimum shelter provisions during lambing period especially when it occurs in winter months. Often, the flocks are penned in the open during fair weather and some permanent/temporary shelters are made used in winter season. Expensive concrete buildings or use of high quality timber for flooring are not required for sheep shelter.

Type of Sheep Housing

There are many different types of housing that can be used for sheep and some of them are as follows:

- Loose housing/open pens is more advantageous as compared to conventional/stall-fed sheds because it is suitable for semi-arid regions and large-sized flocks, it involves less expenditure, it provides more comfort to the animals, it is less labour-intensive, it provides freedom of movement and gives the benefit of exercise.
- A "lean-to" type of shed, located against the side of an existing building, is the cheapest form of building.
- Stilted housing is common in areas with heavy rainfall. Sheds with slated floor pays heavily in return as around many important diseases are largely prevented by better hygiene.

House may be of concrete or kutcha (made of mud or wood) and it largely depends on economic condition of farmer/breeder. Concrete house has a long life, is costlier, and gives less insulation. Kutcha house has a shorter life, is less costly and gives good insulation. Compartments of sheep farm include ewe shed, ram shed, lambing shed, lamb shed and weaner shed, sick animal shed, shepherds rooms, shearing shed and store room (Table 1).

Table 1 : Dimensions of different sheds in a sheep farm

S No	Name of the shed	Dimension LxWxH (m)	No. of animals Housed	Remarks
1	Ewe shed	15 x 4 x 3	60	-
2	Ram shed	4 x 2.5 x 3	8	Make partition length wise
3	Lamb shed	7.5 x 4 x 3	75	Make partition width wise
4	Lambing shed	1.5 x 1.2 x 3	1	Provide manger and waterer
5	Isolation / sick animal shed	3 x 2 x 3	1	Provide proper ventilation and bedding materials
6	Shearing shed	6 x 2.5 x 3	1	Make arrangement for storage of wool
7	Shepherd room	6 x 4 x 3	-	It should be located nearer to flock

Selection of Site for Shed

The shed site should be easily approachable and spacious, dry, elevated, well-drained and protected from strong winds. An East-West orientation ensures cooler environment. When designing a three-sided shelter, the open side should face south away from the prevailing wind. The barn should be easily accessible for deliveries and manure handling. The site should allow for installation of water and electricity.

Floor space requirements

In organized farm, open and closed spaces are kept in the habitat of goats in a scientific manner. In which the open space is usually kept twice as much as the closed space. That is, 1/3 part closed space and 2/3 part open space is kept.Less space is required if sheep are raised on slatted floors or if they have access to an exercise area or pasture. Clean drinking water and feed provided to animals in the habitat.

Recommended floor space requirements for Indian conditions

S No	Age groups	Covered space (sq.m)	Open space (sq.m)
1	Up to 3 months	0.2-0.25	0.4-0.5
2	3 months to 6 months	0.5-0.75	1.0-1.5
3	6 months to 12 months	0.75-1.0	1.5-2.0
4	Adult animal	1.5	3.0
5	Male, Pregnant or lactating ewe/ doe	1.5-2.0	3.0- 4.0

Feeding and watering space requirement

Type of Animal	Space per Animal (cm)	Width of manger/ water trough (cm)	Depth of manger/ water trough (cm)	Height of inner wall of manger/ water trough (cm)
Sheep and goat	40 – 50	50	30	35
Kid/lamb	30 – 35	50	20	25

General considerations for sheep and goat housing

- Shed should be designed to harvest maximum sunlight and bypass as far as possible wind movement.
- Shed should be located on elevated, well-drained sites. The barn should be easily accessible for deliveries and manure handling.
- Site should allow for installation of water and electricity.
- Shed should have cross ventilation.
- Waterers and feeders should not be soiled by faecal material and regularly disinfected.
- Avoid mixing new lambs with old ones of more than two months of age.
- The sheds should be constructed in an elevated area to prevent water stagnation.
- Fodder trees can be grown around the shed, which acts as a source of feed for the growing goats.
- Sheds should be constructed with proper ventilation.
- Walls of the shed should be free from cracks or holes, while constructing.
- Floors of the shed should be firm and should have the capacity to absorb water. The floors should be constructed in such a way, so that it should be easily cleaned.
- For a comfortable house east-west orientation with generous provision for ventilation /air movement to dry the floor will be suitable in tropical region and north to south orientation in temperate region.
- There is no restriction for the length of the shelter, however breadth of shed should not exceed 12 meter and optimum breadth of shelter is 8 meter.
- Height of ewe should be 2.5 meter and height at ridge should be 3.5 meter.
- The height of chain link used for open space should be 4 feet. The length of the overhang should be 75cm – 1 meter.
- **Mangers** should be preferably made of cement concrete with 2 compartments for providing feed and hay at a height of 45-60cm above ground. Size varies depending on number of animals.

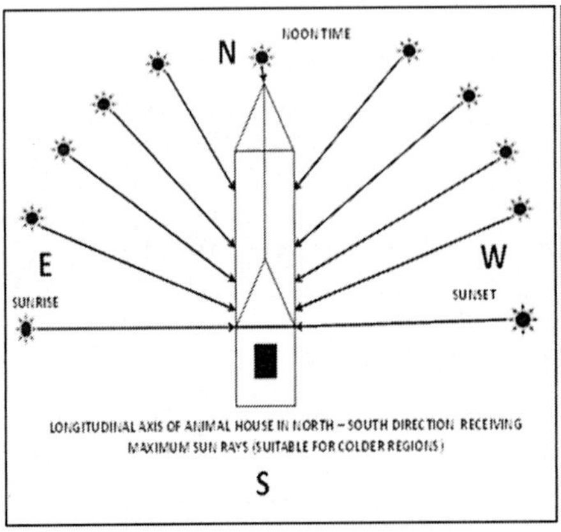

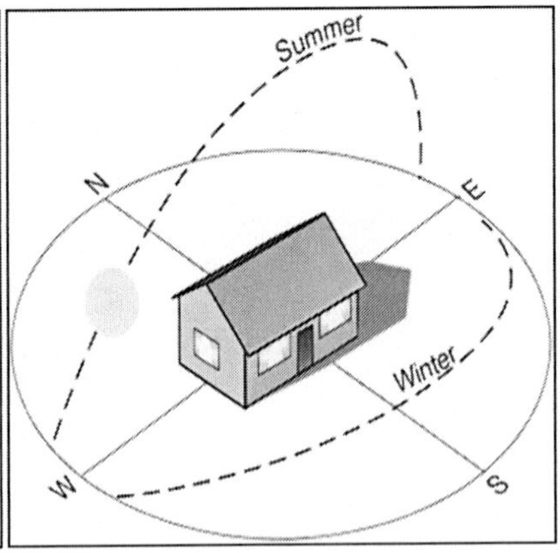

- **Waterers**should be cement concrete or made of galvanized steel. Size varies as per number of animals.
- **Foot bath** should be well embedded in soil and should be located at the entrance to yard to protect animals from foot rot.
- **Dipping tank s**hould be located at one side of yard and can be made of brick in cement.

Construction material in houses

Brick Houses (traditional)

Advantages

1. Structures are usually permanent
2. Not easily demolished by animals

Brick sheep house

Disadvantages

1. It is expensive
2. Houses are often of poor ventilation,
3. Floor needs special attention and may not be easily cleaned.

This system is adopted by sheep breeders who have sufficient money and material with them.

Slated/button floor

Low cost housing for Sheep

Tin Sheds

In this the houses is made of GI tin sheets. It is also low cost housing. The GI sheets must be fixed in strong timber or in anle iron for proper protection. Ventilation may be compromised if proper number of windows are not kept.

Houses of wire mesh wall and GI sheet roof

In this system the side walls are made of wire mesh and roof is made of GI sheet supported by Angle iron. It is cost effective. It has got more ventilation. For new born lambs separate lambing room is made. It can be used intialy when farmers don't have enough money. It can be added with wind shields.

Tin shed

Wire mesh sheep house

Wire mesh house with fibre based shades

References

1. Hand book of Animal Husbandry. V.K. Taneja, C. Chakravarty, C.S. Viswanath, Aruna T. Kumar. Third revised. Indian Council of Agricultural Research, New Delhi.
2. Livestock Production and Management. N.S.R. Sastry and C.K. Thomas. Fourth revised. Kalyani Publishers, Ludhiana, India.
3. Merck Veterinary Manual Online, 2011. Merk and Co

4 Pregnancy Diagnosis, Management of Pregnent Ewes and New Born Lambs

Asloob Ahmed Malik

Pregnancy Diagnosis

Pregnancy diagnosis is a common management practice to identify pregnant and non-pregnant ruminants.Early and accurate pregnancy diagnosis is of considerable significance in improving efficiency of production. It may be important for undertake rebreeding of empty ewes and for proper management of pregnant ones. Number of methods are available for diagnosis of pregnancyin small ruminants and the methods of choice will depend on equipment available, number of days post breeding on which examination is made and the accuracy.

The traditional methods of pregnancy diagnosis are presumptive based on the observation of the animal and includes:

- Non-return to estrus/cessation of heat.
- Abdominal enlargement/ballottement of belly from three months onward.
- Udder development/onset of lactation as suggestive sign of pregnancy.

Other methods of pregnancy diagnosis:

- Recto-abdominal palpation
- Chemical test (Barium Chloride Test):
- Ultrasonography
- Radiography
- Laparoscopy

Care of Pregnant Ewes

- House pregnant ewes in separate enclosures and protect from inclement weather.
- Pregnant ewes should not be crowded together.

- Do not handle the pregnant ewes too frequently and should not be subjected to fright.
- Feed balanced ration with essential minerals and vitamins during winter.
- During third trimester growth of fetus increases, so provide extra allowances.
- Don't take pregnant ewes for long distance grazing.
- Avoid dipping ewes in advanced stage of pregnancy.
- Avoid high temperature in sheds of pregnant ewes as it may lead to birth of immature and weak lambs.
- During pregnancy, ewes should gain body weight of 4-6 kgs depending on the breed.

Care at Lambing

- The gestation period in sheep ranges from 142-150 (Avg. 147) days.
- An ewe that is about to lambprefers to leave the flock to find a sheltered place. She is restless, the udder is often distended and external genitals are in a hyperaemic and oedematous condition.
- Normally there is no necessity of assisting the ewes at the time of lambing except in the case of dystocia.
- Before expected date of lambing, keep ready lambing kit that may include tincture iodine, gloves and towel to assist delivery, I/V infusion set, dexona, glucose and rintose solution.
- Besides arrangements for heating hypothermic lambs or keeping lambs in warm condition for first week post lambing should be made well in advance. In addition, the artificial milk or colostrum should be kept ready.
- Leave the ewes alone to lamb and watch for about half an hour.
- If difficulty is noticed due to failure of normal birth after an hour of straining, the ewe should be rendered assistance under complete asepstic conditions including pulling the lamb provided lamb is in right position.
- Remove mucous from nostrils, mouth and throat of the newborn if needed by gentle gravitation method by hanging the newborn on its hind limbs.
- Allow the mother to lick lamb as soon as possible or make her to do so, if required by sprinkling salt or wheat bran or both on the body of the lamb and if needed dry the lamb with a towel.
- Apply strong tincture of iodine to the navel, repeat twice/thrice on first day and thereafter daily for a few days to avoid navel ill.

- Make sure kids get adequate colostrum within 24 hours.Assist the lamb to suckle colostrum if needed and nurse the orphans or lambs of dams with no milk.
- Feed week lambs/kids usingesophageal tube.
- Record birth weight of lamb properly.
- Help the ewe to own her lamb and to keep it near her especially in primipara (first lambing) to avoid disowning of the lamb.

Care of Ewe after lambing

- Dispose off after births (placenta) properly.
- Wash udder properly after lambing with antiseptic solution.
- Sealed orifices of teats be cleared by removing greasy plug by hand.
- Provide separate clean area in night enclosures.
- Feed lightly on hay or fresh forages with a little amount of grains.
- Provide free access to water at all times.

Raising orphan lambs

To raise lamb whose dam has died one may:

(i) Adopt strange lamb by foster mother

The measures to be taken are as follows:

1. Rubbing foster mother's milk on the rump of lamb and nose of the ewe.
2. Take ewe (s) that has lost their lambs to orphan lamb in a small pen.
3. Putting the whole skin of the dead lamb over the body of orphan lamb.

(ii) Raise orphan lambs by hand

1. Feed small quantity of milk frequently i.e. 30 g milk at 2 hours interval for two days.
2. Increase the quantity of milk with age and weight/ size of lamb and its appetite.
3. Increase the interval between feedings.
4. Make use of nipples and clean bottle.

Raising Orphan Lambs

The following preparation may be tried as colostrum replacement.

400ml……Cow's milk, Egg……1, Sugar…1tsp, Olive oil…1tsp

Weaning

Weaning is separating ewes from lambs so that mothers come to early heat for next lambing and lambs grow and develop better independently. Although it is generally not practized here but as part of scientific management it is considered very important. Weaning is a challenge to any breeder but its benefits in terms of increasing lambing percentage and growth of lambs far outweigh increased managements costs for separate ewes and lambs. Government farms in our state have been practising it routinely. Breeders are advised to consult qualified veterinarian for details regarding weaning and creep feeding. As weaning may initially present a challenge to a breeder but keeping in view great benefits acquired from it, those interested in scientific sheep raising should consider it.

Some important points are listed here.

- Wean ewes from lambs, not lambs from the ewes. If possible, remove ewes from pen out of sight and sound of lambs. If lambs have to be moved to new quarters, leave a couple of ewes with them for a few days to lead them to feed and water locations.
- Lambs should be weaned as early as possible. In our conditions weaning age may be around 2-3 months and earlier the better. Better management is indicated by earlier weaning. The advantage of early weaning is that the ewe's milk production drops off to almost nothing after eight weeks of lactation and may come to next heat earlier.
- Grain should be removed from the ewes diet at least one week prior to weaning and low quality roughage should be fed. Restriction of hay and water to the ewe following weaning lessens the chance of mastitis to occur. Poorer quality roughage should be fed to the ewes for at least 10 to 14 days following weaning.
- Handle the ewes to limited extent for about 10 days following weaning. Tight udders bruise easily. If possible, bed the area where the ewes will rest heavily with straw to form a soft bed for the ewes to lay on.
- Leave the ewes in dry lot on the low-quality hay or straw until their udders have started to dry up and recede. This can take from four to seven days. Do not turn ewes onto pasture immediately after removing them from the lambs. Wait until they are sufficiently dry.
- Do not abruptly change a lamb's ration for two weeks before or after weaning.

References

1. Hafez E.S.E., Hafez B. Reproduction in Farm Animals. seventh ed. Lippincott, Williams & Wilkins; Philadelphia.
2. A Textbook of Animal Husbandry by G.C. Banerjee. Eightth Edition.
3. Applied Veterinary Gynaecology And Obstetrics by Pradeep Kumar.
4. Sheep Management Practices by Ravindra Reddy and Sarjan Rao.

5 Feeding Management of Sheep

Ajaz Ahmad Ganie, Mir Nadeem Hassan, Manzoor Ahmad Yatoo, Iqra Hussain Shah

Sheep generally thrive well on the pasture. Attention should be paid on pasture improvement and management. Rotational grazing should be followed to avoid worm infection and unthriftiness and to ensure the availability of good pasture all the time. The fodder should be conserved in the form of hay and silage for the lean period.

- Water:The sheep should be watered at least one a day at the rate of 2-3 liters per head per day. The requirement of water for the crossbreds during the summer months will be slightly more (5-6 liters). The younger ones may require 1-2 liters of water every day.
- No need of warming water in winter but don't give frozen water.
- Never give water immediately after feeding grains.
- Change the diet gradually for two weeks to avoid diarrhea, acidosis and other problems.
- Give Creep Ration to lambs for quick growth.
- Give flushing (increased quantity of concentrates) to ewes and rams two weeks before breeding
- Give challenge ration before and after two weeks of delivery. Animals need 0.5 times more concentrate during this period.
- Use feed blocks to reduce cost on feed.
- Breeders have tried various things from kitchen waste to soybean, apple pomace, mulberry leaves for optimal utilization of available feed/fodder resources with good results. You should consult a qualified veterinarian for planning best utilization of available feed/fodder resources.
- Group weak animals for separate feeding. Monitor body weight as an indicator of animals health.
- Ground limestone, egg shells etc. can be used as source of calcium.

- Inadequate feeding to ewe results in huge losses in the form of
  1. Shortened gestation.
  2. Reduced birth weight in lambs.
  3. Reduced vigour or rate of progress of the lambs towards suckling.
  4. Reduced thermal insulation of the birth coat and reduced energy reserves of the newborn.
  5. Pregnancy toxemia in ewes.
- The lambs from deficient ewes may suffer from hypothermia and other deficiency syndromes viz. rickets, goiter, swayback, staggers, blindness. These effects are particularly marked in twin bearing ewes.
- Overfeeding in late pregnancy can produce disproportionately large sized lambs and consequently a high incidence of dystocias particularly in maiden ewes.

Proper nutritional management is essential to exploit the full genetic potential of the cross bred sheep. Feed accounts for 55-60% in the total cost of rearing the sheep. Nutritional requirement of sheep at different ages is given below

1. Feeding of preweaned lambs from birth to 90 days of age

The most critical peroid in the life of the lam is first 48 hours. If a lamb is unable to nurse within half an hour after birth, it should be assisted to suckle to get the advantage of colostrum @ 1/10th of the body weight normally at three times a day. The normal birth weight of lamb/kid is 2-3 kg and from 3 days onwards feed milk at 1/10th of the body weight.

Advantages of colostrums feeding

i. Immunity

ii. It is good source of vitamin A and Vitamin D along with minerals like copper, iron, manganese, magnesium etc.

iii. Laxative in nature and help in easy passage of meconium (first faeces)

a. **Creep feeding:** The creep is small enclosure in a sheep pen having opening wide enough for lambs to pass in while ewes are kept back. The practice of providing supplement feed to nursing lambs is called creep feeding. The creep feeding is done from 10 days of age to 90 days to promote growth during early age feeding and rumen development.

Feeding schedule for lambs:

Age	Milk (ml/day)	Creep mixture (g)	Green fodder (g)
0- 3 days	1/10th of B.Wt (3 times /day)		
4-30 days	1/10th of B.Wt (3 times /day)	Small quantity	Small quantity (started at 15 days)
31-60 days	1/10th of B.Wt (2 times /day)	50 – 150	
61-90 days	1/10th of B.Wt (2 times /day)	150 – 200	

Feeding schedule for kids:

Age	Live weight (kg)	Milk (ml/day)	Creep mixture (g)	Green fodder (g)
0- 3 days	2- 2.5	400 ml		
3-15 days	2.5- 3.5	400 ml		
16-30 days	3.5 – 4	500 ml	Small quantity	Small quantity
31-60 days	4 – 6	600 ml	50 – 150	50 – 100
60-75 days	6 – 7	400 ml	100 – 150	250 – 300
75-90 days	7 – 8	200 ml	150 – 200	250 – 300

b. Composition of creep feed

Maize = 40%

Groundnut Cake = 30%

Wheat Bran = 10%

De-oiled Rice Bran = 13%

Molasses = 5%

Mineral Mixture = 2%

Salt = 1%

(Fortified with vitamin B1, B, D and antibiotic feed supplement)

The creep feed should have DCP = 17.4% and TDN = 73%

The weaning should be done at 90 days of age and after the average body weight is around 14 kg.

2. Feeding of growing lambs (finishing lambs/kids from growing to slaughter)

Extensive system will not meet the requirement of growing finishing lambs. Hence should be given concentrate mixture to supplement the nutrients obtained through grazing.

a. Composition of concentrate mixture:

Maize	= 40%
Maize	= 25%
Wheat bran	= 32%
Gram waste	= 26%
Grounnut cake	= 15%
Mineral mixture	= 1%
Salt	= 1%

OR

Maize / Jowar	= 30%
Groundnut cake	= 20%
De-oiled rice bran	= 40%
Molasses	= 8%
Mineral mixture	= 1%
Salt	= 1%

The concentrate mixture should contain DCP = 13% and TDN =65%-70%

b. Feeding shedule:

Body weight	**Concentrate mixture (g)**	**Roughage (g/day)**
12-15 kg	200	400
16- 25 kg	250	600
25 – 35 kg	300	700

Note: - 8 hour grazing can be substituted in place of roughage.

Cross bred sheep should attain more than 30 kg of body weight by 6 months of age.

Thumb Rules of Ration Formulation

The following broad points may be easily remembered for economic and scientific ration formulation.

Wheat bran	25% rich in phosphorus
Rice bran	25% rich in calcium (so Calcium and Phosphorus ratio is maintained and helps to prevent stone formation in urinary tract in males)
Maize	25%
Oil cakes	20% (30% for growers)
Urea	2%
Mineral Mixt.	2%
Salt	1%
Hay	1.5kg daily

Creep Ration

Maize 50%

Oil Cakes 30%

Rest molasses etc. not bran

Sheep require daily intake of around 3% of their body weight

Sheep require daily intake of around 5 litres of water

Feeding of lactating/suckling ewe:- After lambing proper care should be taken while feeding mother ewes, only little grain mixture should be fed during first 2-3 days and ewe should be given good quality forage. Dry matter consumption may go upto 4% of body weight. Lactating sheep require twice the maintenance requirement during first two months of lactation followed by 1.5 times the maintenance during remaining period if twins are born or additional 200g concentrate is given per kg milk produced. The maintenance requirement of concentrate diet for sheep is 250g/day. Feeding of supplemental concentrate mixture can gradually be diminished after 8-10 weeks of lambing and can be stopped after 12-13 weeks of lambing, as lambs are weaned from mother the ewes may be maintained on grazing alone and continue until the next breeding season.

S.No	Stage of lactation	Concentrate diet (g/day)	Roughage
	0-2 months of lactation	400-500 {(250g Maintenance) + (200-250lactation)}	8 hour grazing
	2-3 months of lactation	375g	8 hour grazing

Feeding of pregnant ewes during last six weeks of gestation: During this stage the ewes need more energy, protein, minerals & vitamins to meet the increased requirements for feotal growth and development of the potential for high milk production. Pregnant ewe cannot consume sufficient bulk roughage as the space in the abdomen is reduced due to growth of feotus & its membranes; it has to be fed with good quality forage (cereal fodder 6% DCP and 60% TDN) & a concentrate mixture supplement @ 500g/day. Before 5-7 days before kidding, amount of concentrates should be reduced to about 100-150g/day and laxative diets comprising of soaked bran, leguminous green fodder, hay etc should be given in ample quantity.

Practical way to find out whether the ewe in advanced pregnancy are receiving their quota is to check the average daily gain of 100g for smaller breed & 150g for larger breeds is a fair measure of nutrient intake status.

Feeding requirement for breeding Rams

Sufficient green fodder (more than 8 hr grazing) and concentrate mixture @ 500g/day must be given during breeding season.

Notes

- Early in the creep feeding period, stimulating intake is of primary concern.
- These diets should be formulated to contain upto 20% crude protein.

References

1. Advance Animal Nutrition by D.V. Reddy.
2. Small Ruminant: Production Medicine and Management (Sheep and Goats) by Cody W Faerber and Lyle G McNeal
3. Hand book of Animal Husbandry. V.K. Taneja, C. Chakravarty, C.S. Viswanath, Aruna T. Kumar. Third revised. Indian Council of Agricultural Research, New Delhi.

6 Factors Affecting Increased Fertility and Overall Performance of Sheep in Light of Genetics

Shabir Mir, Aijaz Sydain, Nusrat Nabi, Majid Shafi, Saba Bukhari

Terminology

Crossbred:	An animal whose sire (father) and dam (mother) are of different breeds or breed types.
Purebred:	An animal whose parents are of the same breed or type.
Registered or Pedigreed animal:	An animal with a known ancestry.
Hybrid Vigour:	Natural phenomenon whereby the performance of the crossbred offspring is superior to the average performance of the parent breeds. It contributes to superiority of crossbred animals.
Inbreeding:	Mating closely related animals like ewe's progeny with itself.
Outbreeding:	Mating between unrelated individuals.
Close breeding	More intensive form of inbreeding .eg: Parent-offspring and Full sib mating.
Line breeding	Less intensive form of inbreeding. Eg: Half-sib mating, Cousin mating and Grand parent-grand son/daughter mating.
Outcrossing	Form of outbreeding where mating occur between unrelated animals of same breed.
Grading-up	Form of outbreeding in which mating of the pure-bred sire of the descriptive breed with a local female , generation after generation.
Top crossing	Form of outbreeding where mating of pure-bred male with unrelated female(only for one generation unlike grading-up).
Cross breeding	Mating between animals of different breeds within same species.
Species hybridization	Mating between animals of different species.

Unless the objective is to raise and market purebred and/or registered sheep, it is better to raise crossbred sheep. This is especially true for beginners. It's better to practice shepherding with hardier, less expensive animals.

- Sheep are the seasonal breeders with extended breeding season and can lamb twice a year in some cases. Ewes should be bred twice a year or at least thrice in two years. It is often possible and profitable to get 3 lambs in two years or 5 lambs in 3 years by adopting competent and felicitous mating practices keeping in view the local prevailing conditions.
- Ram used for mating should be changed after every two years to avoid ill effects of inbreeding (inbreeding depression), which can lead to significant reduction in performance of the progeny with deleterious consequences for populations and breed persistence. Moreover the growth rate of new progeny will decrease manifold and congenital problems will increase.
- Rams not used for breeding purpose should be castrated in order to prevent unwanted breeding thus maintaining quality progeny. This practice can also result in improved meat quality and to make management and handling easier and safer.
- One ram can very effectively service 30-40 ewes in a breeding season.
- Best ram should always be selected for the stock. It is always a good investment as a ram is considered to be half the flock.
- Age of the ram should be 2 years or more for breeding purposes.
- Separate the ram from flock for 12 hrs before introducing it to the female pen.
- Usually the ram should be turned to flock at night.
- In temperate climates, most sheep are seasonal breeders and mate when the day length is shorter and lamb in late winter or spring. To produce "out-of-season" lambs we need to select a breed that is able to breed at different times of the year. Many sheep breeds including Merino, Dorset and Roumbouillet from which crosses in Kashmir have been obtained, have extended breeding season and are not very strict seasonal breeders. Many progressive breeders are getting more than one crop from a significant number of ewes annually.
- Any flock of sheep can be selected and managed to minimize care. Progressive breeders may often prefer easy-care traits (e.g. unassisted lambing, minimal hoof trimming and minimal de-worming) in their management and selection program.
- For ram breed selection, a ram with the appropriate type of wool/coat and reproductive characteristics should be selected. Pigmented rams should be avoided for breeding purposes. It should further be ensured that the selected rams have correct legs and pasterns and checked for any lameness.

How to select for increased fertility?

- Improving genetics within a flock starts with selecting a quality ram that complements existing ewe traits and improves any weaknesses in the flock.
- Identify lambs that were born as twins or triplets and select replacement lambs from this group. Twin lambs from young ewes have a greater potential for twinning than do twins from older ewes.
- If additional replacements are required, select single ewe lambs from young ewes.
- Select rams that were twins or from ewes that had high twinning records.
- Ewe lambs that exhibit estrus are typically more fertile and have a greater lifetime production of lambs than ewe lambs that do not reach puberty the first year. Selecting early maturing ewe lambs also emphasizes early season lambing, which might be advantageous in some management systems.
- Productive life of ewes is 6-7 yrs.
- When selection is to be done on visual appraisal, it is important to know all the external parts of the sheep with each part having an ideal characteristic for breeding rams:
- A long, square, level rump
- Long, smooth thick neck
- Adequate diameter of cannon bone
- A long wide loin
- Well sprung ribs
- Deep fore-rib
- Smooth shoulders
- A masculine neck and head
- A trim neat throat and breast
- Strong upright pasterns
- Adequate and evenly sized testicles
- And feet and legs should be placed directly underneath the animal.

Summary

Every sheep operation should invest some time developing a better ram selection strategy. Analyzing the ewe flock will help to determine what traits would be important in ram. Developing a good strategy will yield rams that inturn produce lambs with better performance and daughters that improve genetics within the flock.

References

1. Sheep Production - With Information on the Breeding, Care and Management of Sheep" by W C Skelley.
2. Livestock Production and Management. N.S.R. Sastry and C.K. Thomas. Fourth revised. Kalyani Publishers, Ludhiana, India.
3. Livestock Production Management by C K Thomas and N S R Shastry. Shastry Thomas.

7 Care and Management of Different Age Groups of Sheep

Iqra Hussain Shah, Mir Nadeem Hassan, Muzamil Abdullah, Showkat Ahmad Shah, Ajaz Ahmad Ganie

Sheep represent an important component of human living with regard to food security and employment generation through production of mutton, wool manure, and skin. Management holds the key in getting maximum output from these animals with minimum possible inputs and at the same time ensures animal welfare. A healthy animal in a comfort zoon will certainly perform better in comparison to stressed animals. Proper management is directed towards identifying factors which may be stressful to the animal so that they can be removed in time. However, the comfort needs of the animal often change to some extent with regard to season and physiological status of the animal. Keeping in view the changing needs of the animals, general managemental tips for sheep belonging to different age/ physiological states are given below:

Care and management of breeding animals

I. *Selecting the animals*

Sheep is a seasonal breeder which breeds from September to November under temperate climate of J and K. Proper and timely breeding certainly means better production and profitability.

- Ensure the animal to be used for breeding have attained the proper age and about 70% of their adult weight
- Look for any defects in the animals which may render them unfit for breeding. For example a ram (breeding male sheep) with poorly developed testicles may not be ideal for breeding. Similarly an ewe with poorly developed udder may not produce enough milk for its newborn.
- Both weak as well as fat animals should not be put to breeding as the outcome may not be good
- A breeding ram may be be used to mate around 20-25 ewes.

II. *Flushing the Ewes*

The breeding ewes in particular must have access to good quality feed/fodder during breeding season. This can be achieved by providing them extra concentrate in addition to good quality green and dry roughage starting from 4-6 weeks before the expected breeding time.

III. *Watching for estrous signs*

The ewes need to be watched for following signs during breeding season so as no to miss animals in heat

i. Redding of the vulva and discharge from vulva
ii. Tail wagging
iii. Mounting other animal
iv. Seeking male
v. Frequent bleating
vi. Push her back
vii. Standing for mating (standing reflex)
viii. **Marking the rams:** For proper breeding records, it is recommended to paint the males with some easily identifiable paint like lamp black in their neck region to indicate if a particular ewe has been mated as some of the paint will stick to it while being mounted.

Care and management of pregnant and lactating ewe

Careful management of the pregnant ewes will have a marked influence on the percentage of lambs dropped. Following steps may be taken to afford proper attention to these animals.

i. Do not handle the pregnant ewes too frequently.
ii. Separate the advanced pregnant ewes from the main flock and take effective care in their feeding and management.
iii. Extra feed during the latter part of pregnancy (3 to 4 weeks before parturition) will be beneficial for the condition of the pre-parturient ewes which will help in improving milk production of ewes, and birth weight and growth of lambs.
iv. Prepare and disinfect the lambing pens in advance and ensure provision of dry bedding, proper ventilation, heating arrangement besides clean water.
v. Provide atleast a floor space of 1.5 square meter or 15 square feet for each ewe-lamb pair in the lambing pen

vi. Inadequate and poor nutrition may result in pregnancy toxaemia, abortions, and premature births of weak lambs.

vii. Watch gestation length which ranges from 142 to over 150 days. Early maturing breeds have a slightly shorter gestation period.

viii. Save parturient ewes from cold and chilly weather

ix. Ensure vaccination of ewes with multicomponent clostridial vaccine (MCC) one month before expected date of lambing to provide passive immunity to lambs against lamb dysentery and other clostridial diseases

Care and management of ewes approaching parturition

Proper monitoring of animals during this period is of utmost importance so that timely interventions can be taken up in case of any exigency. Following are some of the important considerations during this period:

I. **Observing for signs of approaching parturition**. These include the following
   - Restlessness with frequent sitting and standing
   - Tendency to stay lonely and leave if flock if the animal is with the flock
   - Swelling and distention of udder
   - Genitals are in a flushed and flaccid condition. Generally in healthy ewe parturition is normal. Still, the following precautions may be taken during and after parturition.

II. **Being ready for assistance:**
   - Proper observation of the ewes will ensure that the assistance can be provided on time, if needed. A solution of povidone iodine, thread to ligating umbilical cord, a pair of sterile hand gloves are some of the basic items to be kept ready in the lambing pen
   - Seek prompt veterinary aid and advice from an experienced shepherd or stockman in case of dystocia

III. **Feeding management:**
   - As lambing time approaches or immediately after lambing the grain allowance should be reduced but good quality dry roughage should be fed free choice
   - It is usually preferable to feed lightly on the day of parturition, but allow plenty of clean, cool water.

Care and management of Lactating ewe

- Soon after lambing the ewe must be given just enough of slightly warm water.
- After parturition the ration of the ewe may be gradually increased so that she receives the full ration in divided doses 6-7 times in a day.

- Bulky and laxative feedstuffs may be included in the ration during the first few days.A mixture of wheat bran and barely or oats or maize at 1: 1 proportion is excellent
- Feed a sufficient quantity of good-quality hay and concentrates (if possible) to the lactating ewes for meeting nutritional requirements for early lactation.
- Provide plenty of clean fresh drinking water as the Lactating ewes drink a surprisingly higher amount of water during lactation.
- If access to good quality greens/6-8 hrs daily grazing is available, concentrate supplementation may not be needed.

Care and management of new born lamb

- In Kashmir regions, lambing season coincides with winter necessitating the heating arrangements for the lamb in the lambing pen well in advance
- Ligate, sever and antiseptically dress the navel cord of the lamb.Povidone iodine may be used for the purpose
- Wipe the nose and mouth clean of mucous. If the lamb does not breathe properly, put it upside down and try to stimulate breathing by rubbing it.
- If it appears lifeless, swing it by its hind legs.
- Allow the lamb to be licked and recognized by its mother properly
- Provide assistance if the lamb is not able to stand up
- Ensure the first faeces (meconium) is shed by the lamb. Give a teaspoon full of castor-oil or liquid paraffin to the lamb to facilitate defecation and passing out of meconium easily
- Colostrum feeding is must for the lamb and more beneficial if fed earlyat the rate of 100 ml per kg live weight on thumb rule basis
- Resort to artificial milk feeding or arrange foster mothers to disowned or orphan lambs. Goats can serve as an excellent foster mother, but ewes that have lost their iamb early after birth may also be utilized
- Unnecessary and frequent handling of the lamb soon after birth is not a good practice and should be avoided
- Allow newborn lambs to be with their mothers all the 24 hours for the first week or so
- Provide creep feedto suckling lambs in addition to the suckling of milk from the tenth day to weaning age
- Newborn lambs are very sensitive to cold and may develop hypothermia if not properly taken care of especially during the first few days after birth. Use

electric convectors, warm blankets and dry bedding. Seek veterinary help if the need arises.

- If possible, make available green leguminous fodders or fresh tree leaves to lambs to nibble during the suckling period
- Identification of lambs is must. They may be ear-tagged or tattooed on the ear for identification as early as possible

Weaning and Care of Weaners

Weaning means separation of the young one from its mother to be reared separately. It is stressfull for the lamb and management of weaners plays an important part in good sheep husbandry. The following steps are important in the proper care and management of weaners.

- Weaning should preferably be done at 90 days. However, it may be carried out as early as 60 days if the mother is not producing enough milk for the lamb or some other reason provided alternate feeding arrangement (creep feed/milk replacer etc.) are available well in advance
- Supplementary feeding and good clean pastures for growing weaners should be provided.
- Weaned lambs should be dewormed against gastro-intestinal parasites by the first month, and vaccinated against enterotoxaemia and sheep-pox.
- Weaners should not be grazed on poorand thorny types of pasture since it could cause skin irritation, injury to the eyes, and damage to wool.
- They should be protected against inclement weather and predators

References

1. Livestock Production Management by C K Thomas and N S R Shastry. Shastry Thomas.
2. Goat Production and Health Management by S K Jindal.
3. A Textbook of Animal Husbandry by G.C. Banerjee. Eightth Edition.

8 Important Sheep Diseases and their Management

Showkat Ahmad Shah, Majid Shafi, Syed Ashaq Hussain, Henna Wani, Mir Nadeem Hassan

Disease/Diseases Condition	Sign	Treatment	Control
Acidosis	Anorexia, suspended rumination. Rumen is distended with fluid. Gurgling fluid sound from rumen. Subnormal temperature. Tachycardia.	Correction of ruminal pH by giving sodium bicarbonate/ magnesium oxide. Use systemic alkalinizers viz. Sodium bicarbonate (2.5-5%) solution I/V. Use of oral antibiotics:- Tetracycline, Ampicillin. Rumenotorics to restore motility of rumen.	Avoid: Accidental ingestion of excess amount of cereal grains such as rice, wheat, oat maize. Over feeding of whole or crushed grains. Feeding of left over roti or rice.
Anemia	Visible sign is loss of natural pink colour of mucous membranes as that of eyes.	Correction of cause which is often parasitic. Iron injections.	Preventive timely dosing.
Bloat	Skin on the left side of the animal behind the last rib (flank) appears distended	treated with anti-foaming agents such as cooking oil or mineral oil or a commercial product such as Bloatosil.	Regulate access to grasses and feed stuffs that result in excess gas formation.
Brucellosis	Abortion mainly during advance & late Pregnancy. Retention of placenta. Infertility, orchitis.	There is no foolpro of treatment. However the following is recommended treatment protocol for non pregnantbrucella infected animals Inj. Oxytetracyclin @ 25mg/ kg b.wt I/M every two days for 4 weeks in combination with Inj. Streptomycin @ 20mg/kg b.wt I/M every two days for two weeks. Treatment protocol for Brucella infected pregnant ewes is .Inj. Ciprofloxacin @ 5mg/kg b.wt I/M daily for 4 weeks in combination with Inj. Streptomycin @ 10mg/kg b.wt I/M for 4 weeks.	Slaughter of infected animals. Disinfection of infected premises. Vaccination:- Brucella cotton – 19 strain vaccine @ 5ml s/c

Braxy	Affected animal may die without showing any previous sign. Lambs may be found dead in the morning. The lambs stop eating, leave the flock, go down, grind their teeth, become comatose & finally die. Lesions: - Abomasum shows reddening with hemorrhages& slight necrosis on the surface. Pericardial & peritoneal cavities contain excess of fluid.	No satisfactory treatment is available.	The feeding of sheep should be monitored properly. Avoid letting animal take very cold water or graze during morning hours in winter. Let them for grazing in winter only after first feeding hay and avoid frosty pasture. Vaccine may be given. Formalized culture of *Cl. septicum* is used.
Black Disease	Affected animal may die without showing any previous sign. There is high rise temperature, dullness &recumbency. There is abdominal pain &periorbital oedema.	No effective treatment. Penicillin may be tried.	Dosing against liver flukes. Vaccination.
Dystocia (lambing difficulty)	Lambs cannot be delivered after a reasonable amount of time and effort.	Competent assistance should be sought from a qualified veterinarian.	
Lamb dysentery	Severe dysentery, abdominal pain, spasms & aim less wandering. Small necrotic areas in intestinal mucosa. Enlargement of mesenteric lymph nodes.	Oral & systemic antibiotics may be tried.	Immunization of ewe prior to lambing.
Laminitis	Inflammation of lamina of feet most often results from the excessive intake of grain. Toxaemia or severe infection can also lead to it. Animals may die before the feet become involved. Recovered animals may exhibit permanent lameness.	Treat infection and toxaemia if it is the cause. Use anti-inflammatory drugs.	Prevent access to easily digestible food like floor, rice, *roti,*
Enterotoxaemia (Pulpy kidney disease)	Affected lambs may jump in the air, fall in the ground go into convulsion & die within few minutes. Circling & staggering. Animal may become blind. Sheep may show signs of glycosuria.	Broad spectrum antibiotics like Chlortetracycline.	ETV-D Clostridium welchi type-D vaccine can be used.

Navel ill (Joint ill)	Persistent enlargement & swelling of umbilicus. Purulent discharge from the umbilicus. The affected young become dull depressed, having no interest for suckling disinclined to move. The umbilicus remains moist & oozing of blood stained serous material. There is formation of umbilical abscess. The joints are swollen, hot & painful. Blood stained serofibrinous discharges may drain out of the affected joints.	Penicillin-streptomycin or ampicillin or cephalexin may be used for at least 7 days. Surgical drainage of umbilical content is to be done. Dressing with povidone iodine is to be done.	Dipping navel in tincture iodine immediately after birth, and then twice for 3 days.
Foot Scald	Affects Interdigital Space. Has no foul odour Less dangerous than footrot and much easier to treat. Not contagious	Placing sheep on drier footing and out of mud will alleviate the problems of the disease. Zinc Sulphate Footbath	Frequent change of night resting place in bahaks and cleaning of sheds and maintaining dry footing.
Footrot	Lameness with a characteristic foul odor in one or more feet is the most common symptom	Controlled and/or eradicated by a combination of hoof trimming, vaccination, foot bathing and soaking and culling. Zinc sulfate is considered to be the most effective foot rot treatment.	Separate affected cases and graze/feed them separately. Maintain a footbath at shed entrance. After treatment the sheep should be held in a dry area for at least 24 hours to allow the medicine to take effect.
Bluetongue	Inflammation, swelling, and hemorrhage of the mucous membranes of the mouth, nose, and tongue. Inflammation of the feet also present. Tongue and mucous membranes of the mouth become swollen, hemorrhagic, and may look red or dirty blue in color.	Antibiotics for covering secondary bacterial infection.	It occurs in summer and autumn and can cause abortions later as well. Avoid access of animals to stagnant dirty water, especially during autumn as insects that transmit the disease breed there.
Contagious Ecthyma	Scab formation on the skin, usually around the mouth, nostrils, eyes, mammary gland and vulva. It first appears as tiny red nodules, usually at the junction of the lips.	Treatment is usually unrewarding. The disease will usually run its course in 1 to 4 weeks. Ensure lambs don't starve and assist in feeding them.	CE is transmissible to humans, causing painful sores that may last for several weeks. (orf) – Thoroughly wash exposed skin areas, and then apply a skin antiseptic. – Keep small children away from infected animals.

Peste de Petits Ruminants PPR	Fever, erosions & bran like deposits on tongue, dental pad & inner side of lips, diarrhoea, coughing, nasal discharge, dyspnoea, corneal opacity, abortion	. No specific treatment. Antibiotics for 3-7 days, mouth wash, Astringents, Fluid therapy, and Multivitamins.	Vaccination at 3-4 months of age @ 1ml s/c.
Sheep Pox	Fever, Pustuler scabby lesions mostly on non-wooly/non-hairy parts of the body like udder, inside of thighs, axilla, face, lips, ears, under the tail.	Antibiotics for 3-7 days. Dressing of cutaneous lesions Antiseptic or antibiotic ointments or lotions may be applied, to control secondary bacterial infection.	Vaccination in endemic areas. Separating sick animals from healthy ones to prevent contact.
Pizzle Rot Image Source: Ag. Dept. of Western Cape	Infection in the sheath area of the ram. Severe irritation and ulceration of the skin around the preputial opening.	Pizzle Rot	Trim the wool Treat with astringent lotions or salves Systemic treatment with antibiotics Limit protein in diet to 15% or less
Uterine Prolapse	The womb (uterus) is turned inside out and pushed through the birth canal by abdominal straining. May occur immediately or several days after lambing. It is a life-threatening condition.	The prolapsed uterus is manually forced back into the ewe, by a qualified veterinarian.	Cull animals with repeated history of prolapsed.
Milk fever	A metabolic disease affecting mostly pregnant ewes near term when calcium requirements are the highest	Administer calcium gluconate subcutaneously/ intravenously as per the severity along with other supportive therapies	Adequate dietary supplements of ca in diet.
Pregnancy Toxemia	Metabolic disease that affects ewes during late gestation. It most commonly afflicts thin ewes, overfat ewes, older ewes, and/or ewes carrying multiple fetuses. It is caused by an inadequate intake of energy during late pregnancy, when the majority of fetal growth is occurring.	Increase the blood sugar supply to the body by administering glucose intravenously or propylene glycol or molasses orally. In extreme cases, removal of the fetuses is the only recourse to save the ewe and lambs. Can be prevented by providing adequate energy to ewes during late gestation, usually ½ to 1 lb. of grain per head per day, more for high producing ewes. Adequate feeder space is also necessary to ensure all ewes are able to consume enough feed	Feed molasses during last trimester of gestation.

Hypothermia (chilling)	leading cause of death in neonatal lambs. Characterized by a body temperature of Mild-102°F Moderate- upto 98° F Severe- below 98°F	Wet lambs should be towel dried and supplemented with heat or put in a warming box using dry heat (heat lamp or hair drier). Colostrum should be tube fed at a rate of 20 to 25 ml per pound of body weight. Once rectal temperature is normal, lambs may be returned to their dams. Dexamethosone injections and intravenous glucose are needed in severe cases.	Adequate prevention against cold to lambs in early life. Special warming box be kept ready during lambing season.
Organophosphate poisoning	Salivation, lacrimation, bronchoconstriction, urination, increased GIT motility, muscle tremors, ataxia, generalized tetany, weakness and paresis	Repeated injections of atropine sulphate.	Avoid access to poisonous substances.
Pica	Animals ingest unusual objects like soil or polythene etc.	Give phosphorus injection.	Ensure mineral supplementation of diet.
Coccidosis	Foetid diarrhea containing mucus and flecks of blood Straining often accompanied by painful vocalization may be found. Loss of weight.	Use coccidostat drugs.	Coccidostat drugs. avoid mixing of very young and old lambs.
Urolithiasis	Signs usually start with restlessness and anxiety. Animalss may experience abdominal pain, urine dribbling, distention and rupture of the urethra. They will usually experience a loss of appetite. They may have a humped-up appearance and edema under their belly. They may kick at their belly and strain to urinate. Dribbled urine may be bloody.	Treatment of urinary calculi depends upon the location of the obstruction and could be as simple as snipping off the urethral process to allow calculi at the end of the penis to dislodge. Tranquilizers and antispasmodics may help to naturally dislodge some calculi. In more advanced cases, surgical intervention may be necessary to save valuable animals. Veterinary advice should be sought in this case.	Decrease the intake of phosphorus. Maintain the calcium to phosphorus ratio 2:1 or above. Add ammonium chloride at a level of 0.5 percent of the total diet. Ensure clean, fresh water is always available. Increase salt intake (force feed) up to 4% of the total diet. This causes frequent urination and higher water intake.

Polioencephalo-malacia	PEM may be acute or subacute. Animals with the acute form manifest blindness, recumbency, tonic-clonic seizures, and coma. Those with a longer duration of acute signs have poorer responses to therapy and higher mortality. Animals with the subacute form initially separate from the group, stop eating, and display twitches of the ears and face. The head is held in an elevated position and there is a staggering. As the disease progresses, there is blindness with a diminished response. Head pressing, turning of neck, and grinding of the teeth may be observed. The subacute form of PEM is frequently followed by recovery with only minor neurologic impairment.	Thiamine is given @ 10-20 mg/kg, IM or SC, TID. Initial treatment may be administered IV. Beneficial effects are usually observed within 24 hr and sometimes sooner; however, if there is no initial improvement, treatment should be continued for 3 days. Dexamethasone at a dosage of 1-2 mg/kg, IM or SC.	Dietary supplementation of thiamine at 3-10 mg/kg feed.
Listeriosis	Initially, affected animals are anorectic, depressed, and disoriented. They may propel themselves into corners, lean against stationary objects, or circle toward the affected side. Facial paralysis with a drooping ear, deviated muzzle, flaccid lip, and lowered eyelid often develops on the affected side, as well as profuse salivation; food material often becomes impacted in the cheek due to paralysis of the masticatory muscles. Terminally affected animals fall and, unable to rise, lie on the same side; involuntary running movements are common.	Penicillin G should be given at 44,000 U/kg body wt, IM, daily for 1-2 wk; the first injection should be accompanied by the same dose given IV. Supportive therapy, including fluids and electrolytes, is required for animals having difficulty eating and drinking.	Affected animals should be segregated. If silage is being fed, use of the particular silage should be discontinued on a trial basis. Spoiled silage should be avoided.

	Listeric abortion usually occurs in the last trimester without premonitory signs. Fetuses usually die in utero, but stillbirths and neonatal deaths occur. The abortion rate varies and has been up to 20% in sheep flocks. abortions and encephalitis occur together in the same flock.		
Struck	Abdominal pain, weakness, depression, failure to nurse &hemorrhagic diarrhoea. There is sudden death &hemorrhagic enteritis.	Oral antibiotics like:- Tetracyline, Chlortetracyline, Penicillin.	Vaccination.

References

1. Radostits O.M., Gay C.C., Hinchliff K.W., Constable P.D. *Veterinary Medicine: A Textbook of the Diseases of Cattle, Horses, Sheep, Pigs and Goats.* Tenth ed. WB Saunders; NY: 2007.
2. Merck Veterinary Manual Online, 2011. Merk and Co.
3. D.G., Baird A.N. *Sheep and Goat Medicine.* second ed. Elsevier Saunders; Maryland Heights, MO: 2012.

9 Footrot in Sheep - its Prevention and Control

Shaheen Farooq, Mir Nadeem Hassan, Iqra Hussain Shah

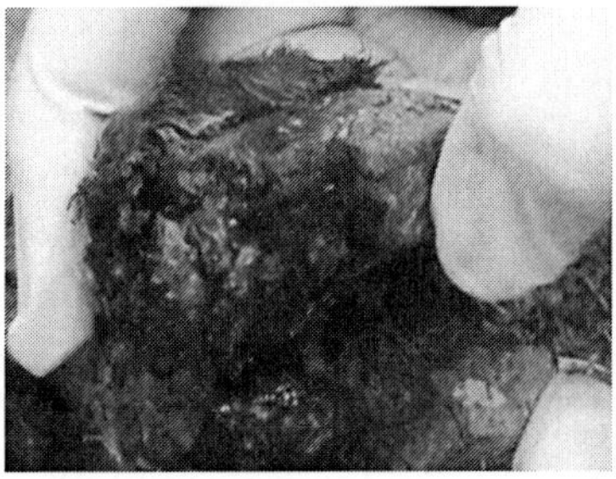

Footrot is an economically important disease particularly of sheep and goats, known to cause significant production loss in addition to cost of recurrent treatment and control. The presence of the disease has not only been established in Jammu and Kashmir but it has also been confirmed from the unusually hot tropical climatic conditions of Andhra Pradesh and Tamilnadu. The disease has also been reported from other parts of the country *viz*: Karnataka, Kerala, Rajasthan, Maharashtra, and Uttrakhand etc. Footrot is characterized by an exudative inflammation with a strongly characteristic foul odour, followed by the necrosis of the epidermal tissues of the interdigital skin and hoof matrix, resulting in separation of the hoof from the underlying soft tissue (Fig-1). The affected sheep exhibit lameness, knee grazing (Fig-2) loss of body condition, reduced wool and meat production and decreased fertility. This debilitating disease is considered to be one of the most important causes oflameness insheepflocks.

Causative agent: *Dichelobacter nodosus* (formerly known as *Bacteroides nodosus*) is the main transmitting agent and *Fusobacterium necrophorum* contributes to the severity and duration of the disease. Both the bacteria are Gram negative and obligate anaerobes. *F. necrophorum* is pleomorphic and non-spore-forming bacterium. It is a normal resident of manure-contaminated environments.

Economic Importance: The disease has been found to generate 10% production loss in body weight and wool growth in affected animals, in addition to recurrent treatment costs and control. In a study conducted by Division of Veterinary Microbiology & Immunology, losses due to footrot were estimated at Rs 198.1 million per annum.

Prevalence of footrot in Jammu &Kashmir

In one of the study conducted by the Division of Veterinary Microbiology & Immunology, the overall prevalence of *D. nodosus* in ovine footrot was 16.19 %

Predisposing factors: Many predisposing factors like the presence of water logged pastures in and around the villages where the sheep are reared make them vulnerable for the persistence of the disease, the sheep flocks in Jammu and Kashmir being migratory, migrate to high land pastures during summers where they are exposed to other predisposing factors like hard rocky surfaces, thorns, wires & sharp objects (which cause injury to feet), muddy pastures, frequent rains and temperate climate which make the disease transmission all the more possible. Also during this migration, flocks from different areas intermingle thereby facilitating the spread of infection from animal to animal and from flock to flock.

Transmission: The bacteria that causes footrot is spread from infected sheep to the ground, manure, bedding, etc., where it is then picked up by non-infected sheep. Footrot is introduced by purchase of an infected animal or by simply using facilities or trucks that have been contaminated by infected sheep. Spread occurs best when temperature ranges from 40-70°F and the environment is wet. Since the organism survive in the environment (all types of soils) for more than 30 days while survival is significantly longer in clay soil, carriers in the flock will continue to reinfect unless the animal is either culled or the organism is eliminated by proper treatment.

Diagnosis

- Footrotis usually diagnosed on the basis of clinical signs, lameness is usually the major sign of an infected animal, although sheep with an early infection may not exhibit lameness. The affected sheep within the flock show limping and lag behind the flock. The area between the toes first becomes moist and reddened. Then the infection invades the sole of the hoof, undermining and causing separation of the hoof from underlying soft tissue. The infection causes a characteristic foul odor and may infect one or more feet at the same time. Not all lame sheep have footrot. Other diseases that may be confused with footrot are foot abscesses, foot scald, laminitis or founder, corns, traumatic injuries, and foreign bodies lodged between the toes.
- Isolation & Identification of D. nodosus: Isolation of the organism is carried out under strict anaerobic conditions. Confirmation of the suspected colonies with ground glass appearance (Fig-4) as D. nodosus (straight or curved rods with characteristic knobbed ends) is carried out by demonstration of the typical cellular morphology in Gram stained smears (Fig-5). Direct detection of organism can be done through Polymerase Chain Reaction (PCR) using 16SrRNA gene specific primers without the need to culture the organism.

The serogrouping of the bacterium is carried out by multiplex PCR (m-PCR) using serogroup-specific (A-I) primers.

1. *Treatment*

It includes :

a) **Foot trimming:**This reduces the number of cracks and crevices where bacteria can hide, removes infected hoof, and exposes the organism to air and various medications. All affected tissue should be trimmed away. Many times, this involves removing a large portion of the hoof wall as well as the overgrown portion. This is necessary if the medication and oxygen are to reach the bacteria to kill them. Foot trimming should be done at least one to two times per year as a part of normal management practices, and more often in conjunction with footbaths in the control of ovine footrot. When trimming feet, it is important to disinfect the trimming instruments (foot shear, hoof parer, or knife) between animals to prevent spreading of the infection. During a severe outbreak, trimming without any other treatment may actually increase the severity of the disease.

b) **Footbaths/Footsoaks:** There are two different types of solutions commonly used in foot baths: zinc sulfate and copper sulfate. For treatment, they should be used 1-2 times per week for several weeks. They may also be used routinely after foot trimming and as a preventative measure.

   - Zinc sulfate (10% solution) is the most effective and least toxic of the two baths. A surfactant or wetting agent (detergent) can also be added to the baths to increase their penetration into the cracks and crevices of the hoof. Use of zinc sulfate or copper sulfate solutions as a foot soak (30-60 minutes of contact) increases their efficacy in a treatment program.
   - Copper sulfate (bluestone) solutions (10% solution) are also useful but are toxic if consumed by the sheep. Hot water helps dissolve it, as does the addition of some vinegar. It is very corrosive so it should not be used in metal foot baths. It also stains the wool.

c) **Dry chemicals:**Zinc sulfate (dry) can be placed in a box in an area sheep must walk through. This will not treat infected animals, but will help decrease the spread of the disease. Lime, disinfectants, or drying agents may be used around feed or water troughs to reduce moisture and decrease the spread of the disease.

d) **Oral therapy:**Zinc sulfate at the rate of 1/2 (0.5) gram per day for 21 days may be helpful both in treatment and prevention, especially if the diet is zinc-deficient. High levels of certain antibiotics may also be helpful in some situations but should only be used after consultation with a veterinarian.

e) **Injection of antibiotics:** While planning antibiotic therapy for control of footrotEnrofloxacin should be considered first. Single injections of long acting tetracycline have also been successful in some cases. The antibiotic therapy should continue for upto one week for complete recovery from disease.Use of any of these antibiotics should be after consultation with or by a veterinarian and should never be used on animals that are intended for slaughter before an adequate withdrawal time is ensured.

2. *Prevention and control*

It is always easier and less expensive to prevent footrot than to treat it after it has established in a particular location. To remain disease free, there are six management principles that will help keep footrot from being introduced into a healthy flock.

a) Never buy sheep with footrot or from a flock infected with footrot, even if the animal(s) appear unaffected.

b) Avoid buying sheep at sale yards or livestock markets where clean and infected sheep may have been comingled or run through the same area.

c) Avoid using facilities (trails, corrals, dipping areas) where infected sheep may have been in the last two weeks.

d) Never transport sheep in a vehicle that has not been properly cleaned and disinfected.

e) Trim and treat the feet of all new arrivals, then re-examine them periodically during the 30-day isolation period.

f) *Vaccination:* Currently in India, there is no effective control measure available in the market for this disease and the farmers suffer unabated economic losses. The commercial vaccines available around the world have limited scope in J&K because of serogroup and strain variation. Keeping in view the colossal economic losses incurred by footrot in the state of J&K in general and Kashmir valley in particular the Division of Veterinary Microbiology & Immunology, Shuhama, SKUAST-K has developed serogroup specific whole cell vaccine against virulent footrot (Fig-6) which is having both prophylactic and satisfactory therapeutic effect. The vaccine is prepared with an oil adjuvant and is given subcutaneously in two doses 30 days apart @ I ml per animal to any age group starting from 3 months of age. The vaccine has protected the animal for about 4 months, but is yet to be commercialized on large scale to meet the field demand.

References

1. Farooq. S, Wani S.A, Hassan M.N., Nazir N., Nyrah Q.J. (2015). "The detection of Dichelobacter nodosus and Fusobacterium necrophorum from ovine footrot in Kashmir, India". Anaerobe 35: 41-43.
2. Farooq S., S .A. Wani, M. N. Hassan, S. Aalamgeer, Z. A. Kashoo, S. N. Magray M. A. Bhat (2018). "The detection and prevalence of leukotoxin gene variant strains of Fusobacterium necrophorum in footrot lesions of sheep in Kashmir, India." Anaerobe 51: 36-41.
3. Farooq, S. Wani, S.A., Hussain I, and Bhat,M.A (2010). "Prevalence of footrot in Kashmir, India and molecular characterization of Dichelobacter nodosus." Indian Journal of Animal Sciences, 80(9):826-830.
4. Shakil A. Wani, S. Farooq, Z. A. Kashoo, I. Hussain, M. A. Bhat, M.A. Rather, S. Aalamgeer (2019). "Determination of prevalence, serological diversity and virulence of Dichelobacter nodosus in ovine footrot with identification of its predominant serotype as a potential vaccine candidate in J&K, India." Tropical Animal Health and Production June 2019, Volume 51, Issue 5, pp 1089–1095,https://doi.org/10.1007/s11250-018-01788-9

10 Importance of Deworming in Sheep & Goat

Zahoor A. Wani, Idrees M. Allaie, Kamal H. Bulbul, Shahana R. Tramboo, Insha Akbar, Showkat Ahmad Shah

Introduction

Gastrointestinal (GI) helminth parasitism isone among themajor constraints for profitable rearing of sheep and goats, particularly under humid environmental conditions.GI parasites includeflukes, tapeworms and round wormsthat occur in digestive and respiratory systems are often represented as sub-clinical infections which cause economic losses in the form of decreased growth, reduced milk, meatand wool production, impair reproductive performance (in females) and lower feed efficienciesand by way of costs incurred on treatment and control.Besides thenegative impacts on productivity, they also make animals susceptible to bacterial, fungal and viral diseases and also steal resources from their hostswhich indicates thatan animal needs more feed to achieve the same amount of growth.Since the occurrence of parasitic diseases is greatly influenced by varying climatological and ecology factors, it therefore implies that the parasitic fauna of different species of livestock mapped out accurately in different seasons of the year forms fundamental information on which prophylactic measures can be based up on.Although several control measures are operated either separately or in combination (integrated parasite management), but chemotherapy and chemoprophylaxis are still the most effective weapon for fighting the menace of parasites.

When to suspect parasites

- Stunted growth
- Prolonged diarrhea
- Anemia with pale mucus membrane of eyes and mouth
- Bottle jaw with normal body temperature
- Low productivity
- Generalized weakness
- Death

Important parasitic Diseases of Sheep & goats

i) **Trematodes** like fasciolosis, immature paramphistomosis and schistosomosis

ii) **Cestodes**like *Moniezia*, *Avitellina*, *Stilesia*, *Thysanosoma* and *Thysaniezia*

iii) **Nematodes** like *Haemonchus*, *Ostertagia*, *Trichostrongylus*, *Nematodirus*, *Bunostomum*, *Oesophagostomum*, *Strongyloides* and *Trichuris*

Deworming

Deworming means giving an anthelmintic drug to an animal to get rid of worms is an important preventative care regime for reducing helminth parasitism, improving animal's health and prevent further reinfection or infestation for a certain period of time. It is also called drenching. Usually only 20% of animals harbor majority of parasites (80%) in a flock or herd. Deworming is categorized as periodic deworming which means deworming the animal on a **regular basis and as per the necessity or strategic deworming that comprises of** deworming the animal as per necessity. The appropriate use of dewormers is critically important as there is chance of development of resistance to dewormers among certain types of worms, making treatment even more difficult.

Ideal properties of anthelmintic/ dewormer

- It should be efficient against all parasitic stages of a particular species.
- It should have broad spectrum of activity against mature and immature parasites including hypobiotic larvae.
- It should be non-toxic to the host, or at least have wide safety margin.
- It should be rapidly metabolized and excreted by the host, otherwise would require long withdrawal periods in meat and milk producing animals.
- It should be easily administered, otherwise they will not be readily accepted by owners.
- It should be compatible with other compounds or drug formulations.
- It should have economic integration and management system
- It should not have any unpleasant side effects.

Deworming of sheep and goat-why?

- Reduce the internal helminth parasite challenge on the grassland.
- Reduce the risk of the negative impact of endo-parasites on sheep and goat.

Using dewormers at proper times of the year (strategic deworming) and/or production cycle when sheep are most vulnerable to parasitism helps in increasing the efficiency of treatment and reduces the overall use of chemicals. This will also decrease cost and help to maintain the effectiveness of the chemical. One of the most important times to deworm is just after lambing, when there is a sudden release of infective eggs within the ewe's intestinal tract. Treating animals at this time minimizes the exposure of larva to newborn as well as young lambs and also reduces pasture contamination. Larval stages of certain nematodes like *Haemonchus* and *Ostertagia*undergo arrested development within the body of host (hypobiosis) at a time when environmental conditions are not favourable for their development. Resumption of the development of these arrested larvae upon return of favourable conditions plays an important role in the survival of these nematodes. Therefore, anthelmintic treatment of animals at the time of resumption of development of these arrested larvae will help in reducing the contamination of grazing places. Larvae may undergo arrested development as a result of acquired immunity and resume development when the existing worm burdens are eliminated or when temporary relaxation in the immunity like late pregnancy and early lactation resulting in periparturient and postparturient rise in faecal egg counts therefore strategic deworming at this particular time will also help in preventing pasture contamination.

Rotating anthelmintics

There are three to four main groups of dewormers/ anthelmintics mainly used by the sheep and goat farmers-

Benzimidazoles (e.g. albendazole, fenbenzadole, triclabendazole);

Imidazothiazoles (e.g. levamisole, tetramisole);

Salicylanilides (closantel, niclosamide, oxyclozanide, rafoxanide); and the avermectins (e.g. ivermectin).

Using a different group of anthelmintic after every 1-2 years will help to keep the parasite resistance under control. Performing fecal egg count of the farm animals will help to determine if resistance to a particular group of anthelmintic is occurring or not.

Deworming schedule

In order to keepanimals in good health and productive, it is necessary to deworm the animals regularly and at the proper time of the year. Based on the extensive epidemiological surveys carried out on parasitic fauna of different species of livestock in different seasons of the year, below mentioned deworming schedule has been framed.

S. No.	Type	Deworming plan
1	Liver flukes	Twice a year in an endemic area of tropical/ subtropical region. Thrice a year in an endemic area of temperate region.
2	Round worms	Thrice a year in an endemic area of tropical/ subtropical region. Twice a year in an endemic area of temperate region.
3	Tapeworms	Twice a year in an endemic area of tropical/ subtropical region Once a year in an endemic area of temperate region

In endemic areas, marshy water-logged lands or where animals are fed paddy straw, dosing on three occasions with broad spectrum anthelmintics on the following dates are recommended for control of flukes in temperate region.For other helminth infections deworming schedule should be followed as:

Worms	Temperate region	Tropical/ subtropical region
Liver flukes	• 15 February to 15 March • 15 June to 15 July • 15 October to 15 Nov	• 1 May to 31 May • 15 November to 15 December
Round worms	• 1 May to 31 May • 1 November to 30 November	• 15 March to 15 April • 15 June to 15 July • 15 October to 15 November
Tapeworms	• 1 November to 30 November	• 1 May to 31 May • 15 November to 15 December

Note

i. A broad spectrum anthelmintic or a drug combination containing flukicide, anti-nematodal and anti-cestodal drugs can be used for common dosing against trematodes, nematodes and cestodes.

ii. Similarly, some drugs which are effective against both nematodes and ectoparasites can be used for common prophylaxis against these parasites.

iii. However, target selective treatment of heavily infected animals in a herd should be adopted after assessing parasitic load as and when needed. This is very important because by adopting this strategy the routine prophylactic treatments can be minimized/ reduced in number.

Dos and Don'ts in Deworming

- Don't overdose or under-dose.
- Go for fecal examination before deworming the animal.
- Judicious use of antiparsitic drugs at proper times of year based on epidemiological studies.
- Change of group of antiparasitic drugs after every 1-2 years, but frequent change at every dosing is not recommended.

- Pregnant animals can be dewormed with pregnancy safe anthelminticsespecially one or two weeks before expected delivery.
- Albendazole and mebendazoleshould not be usedduring the 1stmonth of pregnancy and avoid use of Vit. E/Se- in pregnant ewes along with anthelmintics.
- In lambs, 1stdosing may be done after attaining age 45 days with broad spectrum anthelmintic.

References

1. Soulsby EJL. *Helminths, arthropods and protozoa of domesticated animals.* 7. London: The English Language Book Society; 1982. pp. 126–173.
2. Godara R, Sharma RL. Parasitic infections in livestock at Jaipur. *J Vet Parasitol.* 2010;24:193–195.
3. Gupta JL. *Sheep production and management.* 1. New Delhi: CBS Publishers and Distributors; 2006. pp. 1–239.
4. Shahnawaz M, Shahardar RA, Wani ZA. Seasonal prevalence of platyhelminthosis of sheep in Ganderbal area of Kashmir valley. *J Vet Parasitol.* 2011;25:59–62.

11 Common Parasitic Diseases of Sheep and their Treatment and Control

Idrees M. Allaie, Zahoor A. Wani, Kamal H. Bulbul, Aiman Ashraf, Mufeeda Tariq

Introduction

Parasites or parasitic infections are often represented as sub-clinical or chronic forms in livestock which cause huge economic losses in terms of mortality, morbidity, less production of mutton, wool, decreased fertility and expenses incurred on their treatment and control. In spite of the significant production losses that run into millions of rupees (Shan and Chaudhry 1995), the problem is neglected due to its insidious and chronic nature.Sheep are affected by various parasites like platyhelminthes (trematodes and flatworms), nemathelminthes (round worms), ectoparasites (ticks, lice flies, mites, etc) and protozoa (*Trypanosoma*, *Eimeria*, *Babesia*, *Theileria*, *Toxoplasma* etc). The occurrence of theseparasitic infections depends upon the local climatic conditions and the managemental practices adopted in particular area. Therefore, a thoroughknowledgeabout the parasitic fauna inparticular area should be knownfor the development of suitable control measures.Parasitic infections are usually diagnosed by examining faecal, urine, blood and tissue samples etc. Accurate diagnosis of the parasitic diseases is essential to understand the epidemiology for framing parasite control strategies, which assist in the improving the productivity of sheep in the globe.Although there are different ways of combating parasitism, but each control method in use at present has certain limitations. The general opinion is that strategic use of anthelmintic combined with better managemental practices will help in reducing parasitic burdens to considerable levels which in turn will reduce mortality and production losses to a greater extent. Therefore, this chapter presents a brief overview about different parasitic infections encountered in sheep and their diagnosis and control.

S. No.	Type of Parasite	Name of the Parasite	Organ of Location	Important Clinical Signs/ lesions	Treatment
A) Platyhelminthes					
1. Trematodes (Flukes)					
I		*Fasciola* (*Fasciola hepatica, F. gigantica*)	Liver and Gall bladder	Blood stained froth appears at nostrils and blood discharges from the anus, oedema of mucosae and bottle jaw, pain is felt when pressure is exerted over the liver area, anemia and weakness.	Triclabendazole @ 10mg/kg b.w Oxyclozanide @ 15mg/kg b.w Rafoxanide @ 15mg/kg b.w Closantel @ 7.5mg/kg b.w Nitroxinil @ 7.5mg/kg b.wDiamfenetide@120mg/kg b.w Benzimidazoles like Albendazole, luxabendazole and fenbendazole
II		*Fascioloides magna*	Liver	Extensive liver damage, necrosis of liver along with adhesive peritonitis.	Albendazole @ 15mg/kg b.w Triclabendazole @ 20mg/kg b.w
III		*Dicrocoeliumdenriticum*	Bile duct	Anemia, oedema and emaciation. Cirrhosis of liver and proliferation of glandular epithelium in heavy infections.	Albendazole @ 15mg/kg b.w Fenbendazole @ 150mg/kg b.w Cambendazole @ 25mg/kg b.w Thiabendazole @ 200-300 mg/kg b.w Diamfenetide @ 200 mg/kg b.w
IV		*Eurytremapancreaticum*	Pancreatic duct, rarely in bile duct and duodenum	Clinical signs are not marked but poor body condition may be noticed. Inflammatory foci and granulomata in the walls of ducts is seen.	High doses of albendazole and praziquantel may be tried
V		*Ogmocotyleindica*	All parts of digestive tract behind oesophagus	Clinical signs are not marked.	-
VI		*Paramphistomumcervi, Cotylophoroncotylophorum, Ceylonocotylestreptocoelium, Gastrothylaxcrumenifer, Fischoederius elongatus* and *F. cobboldi.*	Rumen and reticulum	Oedema (sub-mandibular) and emaciation, profuse fluid foetiddiarrhoea and extreme thirst in animals. Adults cause loss of ruminal papillae. Immature amphistomisis result in frank haemorrhagic duodenitis and hypoproteinaemia.	For immature amphistomoisis Oxyclozanide @ 20mg/kg b.w Niclosamide @ 100-150mg/kg b.w Niclofolan @ 4-6 mg/kg b.w Resorental @ 65mg/ kg b.w Brotianide @ 15mg/kg b.w

VI		*Schistosoma bovis, S. japonicum, S. mattheei, S. intercalatum, S. spindale, S. indicum*and *S. nasalis.*	Portal and mesenteric veins	Diarrhea or dysentery, dehydration and anorexia, anaemia and sometimes oedema. Pathological lesions include haemorrhagic lesions on intestinal mucosa and appearance of 'Clay-pipe stem fibrosis' in liver.	Stibophen @ 7.5mg/kg b.w Lucanthone 30-50mg/kg b.w Hycanthone @3-6mg/kg I/M Niridazole @ 100mg/kg Trichlorofon @ 50-70mg/kg b.w
VIII		*Orientobilharziadattai*and *O. turkestanicum.*	Mesenteric and hepatoportal veins	Debility, marked liver cirrhosis and nodules in the intestines.	-
2. Cestodes (Flatworms)					
I		*Moniezia expansa*	Small intestine	Light infections are of little importance, but heavy infections may cause diarrhea and unthriftiness.	Albendazole @5-7.5 mg/kg b.w Fenbendazole @ 5mg/kg Oxfendazole @ 15mg/kg Praziquantel @ 15mg/kg Resorental @ 65mg/kg Bithinol @ 200mg/kg Niclosamide @ 75-150mg/kg
II		*Avitellinacentripunctata*	Small intestine	Symptoms similar to *Moniezia*, but are considered to be less pathogenic.	
III		*Stilesia hepatica*	Bile ducts	Practically non-pathogenic, but occlusion of bile ducts forming sac like dilatations and slight liver cirrhosis may be noticed	
IV		*Stilesiaglobipunctata*	Small intestine	Nodule formation in duodenum and jejunum.	
V		*Thysanosomaactinoides*	Bile, pancreatic ducts and small intestine	Digestive disorders and unthriftiness.	
VI		*Thysanieziagiardi*	Small intestine	No clinical significance	
VII		Metcestodes of *Taenia hydatigena, T. ovis, T. multiceps*	Mesentery, serosal surfaces of organs, skeletal and cardiac muscles, brain and spinal cord.	Clinical signs depend upon the location of the cysts.	-

B) Nemathelminthes (Round worms)					
I	Gastro-intestinal nematodes	*Haemonchus-contortus; Mecistocir-russ*pp.;*Marshal-lagia*spp.;*Oster-tagia*spp.	Abomasum	*Haemonchus* is vigorous blood sucker leading to anaemia and hypoproteinemia. *Mecistocirrus* and *Marshallagia*have similar pathogenesis but is less than *Hae-monchus.Osterta-gia*penetrates gastric glands leading to formation of nodules which give morocco leather appearance.	Fenbendazole @ 5mg/kg; Albendazole @5-7.5mg/kg; Levamisole @ 7.5mg/kg; Morantel tartarate @ 10mg/kg b.w; Phenothiazine @200-400 mg/kg b.w; Ivermectin/ Doramectin/ Moxidectin @ 0.2mg/kg s/c.
		*Trichostrongy-lus*spp.	Abomasum & Small intestine	Villous atrophy and loss of plasma into intestine. Ringworm like lesions due to hyperplastic gastritis.	
		*Coope-ria*spp.,*Nemato-dirus*spp.,*Stron-gyloi-des*spp.,*Bunosto-mum*spp.	Small intestine	*Cooperia* is less pathogenic. *Nemato-dirus*causes villous atrophy and diarrhoea. Penetration of larvae of *Strongyloids*into skin around feet causes dermatitis and sheep become more susceptible to foot rot. *Bunostomum* are hook worms and are active blood suckers.	
		*Oesophagosto-mum*spp.	Caecum and Colon	Results in formation of nodules.	
		Trichuris spp.	Caecum	Diarrhoea, sometimes accompanied with passage of mucus and blood.	
		Chabertiaovina	Colon	Immature adlts and adults are attached to the mucosa and are plug feeders thus highly pathogenic resulting in protein losing enteropathy.	
II	Lung worms	*Dictyocau-lus filaria; Protostrongylus-rufescens.*	Bronchi of sheep and goats	Clinically animals have cough, moderate dyspnoea, fever and loss of condition. At necropsy lesions are exudates in bronchi-oles and scattered patches of consoli-dation.	Fenbendazole @ 5mg/kg; Levamisole @ 7.5mg/kg; Tetramisole @ 15mg/kg b.w; Diethyl carbamazine @ 22mg/kg b.wi/m for 3 days or 50 mg/kg orally for 5 days; Ivermectin @ 0.2mg/kg s/c

		Muelleriuscapillaris	Lung parenchyma of sheep and goats	Infestation of goats leads to a diffuse infection quite different to a nodular reaction in sheep and to the production of an interstitial pneumonia.	
C) **Ectoparasites**					
I	Lice	*Damaliniaovis*	All body parts	Irritation of skin leading to scratching, rubbing, licking of affected part leading to restlessness and damage to fleece and hides. Foot lice may even lead to lameness in sheep.	Insecticides in the form of dust, spray or dip should be used. Ivermectin or doramectin can be given as s/c
		Linognathusovillus, *L. africanus* and *L. pedalis*	Face, legs and feet.		
II	Flies	*Cullicides*, *Simulium*, Sandflies, mosquitoes, *Glossina*, *Lucilia*, *Calliphora*, *Phormia*, *Chrysomyia*, *Oestrusovis*, *Melophagusovinus*etc	Suck blood from any part of the body	Irritation, annoyance, restless, animal becomes unmanageable when in harness. Larvae of some of these flies cause myiasis and *Oestrusovis* larvae in particular causes a condition known as false gid. These also transmit various bacterial, viral and protozoan diseases to sheep.	Affected wound can be treated with larvicidal dressing and antiseptic. Insecticides like Diazinon, cumaphos, fenchlorophos and chlorfenvinphos can be used (0.05-3.0%) Ivermectin @ 200μg/kg and closantel @ 15 mg/kg are also effective Synthetic pyrethroids are also used. Residual spray insecticides can be applied to the walls for killing adult flies.
III	Ticks	*Otobiusmegnini*	Larval and nymphal stages are parasites in the ears	Irritation and inflammation of the ear. Animal appears dull and does not feed well.	Application of insecticides like Diazinon, cumaphos, fenchlorophos, arsenic, butacarb, dioxathion and chlorfenvinphos on the body of animals, living quarters, premises, pastures, etc.
		Various species of *Ixodes*,*Boophilus*,*Hyalomma marginatum*, *Rhipicephalus*,*Haemaphysalis*,*Dermacentor* and *Amblyomma.*	Different body parts	Weakness, anaemia and tick paralysis in animals.	
IV	Mites	*Demodex ovis*	Hair follicles and sebaceous glands.	Usually disease is rare, but scaly form develops.	Topical application of various acaricides Systemic acaricides like dimethoate, trichlorofan, asuntol+neguvon, ivermection or doramectin and closantel

		Psorergatesovis	Skin	Irritation leading to biting and scratching. Wool of the affected areas becomes thready and general appearance is that of a louse infected sheep.	Lime Sulphur dips @ 1% polysulphide Sulphur 0.1% HCH or 0.2% malathion
		Sarcoptesscabiei	Skin (Head) (Non wooly portion)	Disease is rare.	Diazinon @ 0.01% Propetamphos @ 0.0125% Coumaphos @ 0.1% Phoxim @ 0.05% Amitraz @ 0.05% Flumethrin @ 0.05% Ivermectin @ 200µg/kg s/c
		Psoroptes ovis	Body of sheep	Crust and papule formation on the skin (Sheep scab) followed by biting and scratching of the affected part with loss of wool.	Dipping with acaricides like Diazinon @ 0.01% Propetamphos @ 0.0125% Coumaphos @ 0.1% Phoxim @ 0.05% Amitraz @ 0.05% Flumethrin @ 0.055%
		Chorioptesovis(synonym of *C. bovis*)	Pastern region	Scaly condition on the legs and scrotum.	HCH washes are effective Organophosphorus insecticides like fenchlorphos, trichlorofan, or diazinone
D) Protozoa					
I	*Eimeria*	*Eimeria arloingi, E. ahsata, E. ovina, E. gilruthi, E. weybridgensis*and *E. crandalis.*	Small intestinal epithelial cells	Brownish to yellowish green diarrhea streaked with blood, abdominal pain, anaemia, inappetence and loss of weight.	**Curative treatment includes** Sulphadimidine&sulphamethazine 140mg/kg for 3 days Nitrofurazone @ 15mg/kg for 7 days Amprolium @ 50-62.5mg/kg as single dose **Prophylactic treatment in lambs includes** Sulphadimidine&sulphamethazine @ 25mg/kg for 1 week Nitrofurazone @ 0.04% in feed for 21 days Amprolium @ 50mg/kg b.w for 21 days Monensin @ 20mg/kg feed fed continuously Lasalocid @ 25-100mg/kg in feed from weaning

II	*Toxoplasma*	*Toxoplasma gondii*	Variety of cells and tissues.	Infection during early gestation (45-90 days) results in foetal death and expulsion. Infection at 120 days of gestation causes foetal infection but not death and lambs usually survive. Usually mummification, abortion, stillbirth and neonatal death is seen.	Spiramycin, monensin, lasolacid and toltazuril can be tried. Vaccination of ewes with a live attenuated S48 strain (Toxovax)
III	*Neospora*	*Neospora caninum*	Different tissues	Clinical presentations are characterized by abortions, stillbirths, or the delivery of congenitally infected but healthy lambs	Monensin, decoquinate, ponazuril and toltazuril can be used.
IV	*Sarcocystis*	*Sarcocystisarieticanis, S. tenella, S. giganteaand S. medusiformis*	Striated muscles and central nervous system	Gastrointestinal disorders, anaemia, abortion, nervous signs, premature birth and death.	Amprolium and salinomycin may be tried prophylactically.
V	*Babesia*	*Babesia motasi& B. ovis*	Haemoprotozoa; RBC's	Rise in body temperature, anorexia, increase in respiration and heart rate, anaemia and icterus followed by haemoglobinuria.	Diminazene aceturate @ 2-3.5mg/kg by deep i/m; Imidocarb @ 0.5-1mg/kg s/c
VI	*Theileria*	*Theileriahirci*	Haemoprotozoa; Lymph nodes and RBC's.	Enlargement of the lymphnodes, fever, anorexia, dyspnoea, anaemia.	Buparvaquone @ 2.5mg/kg b.w two doses 48 hr apart i/m.
VII	*Anaplasma*	*Anaplasmaovis*	Rickettsial disease; Erythrocytes	Disease is usually sub-clinical.	Tetracyclines
VIII	*Eperythrozoon*	*Eperythrozoonovis*	Rickettsial disease; Infecting surface and periphery of erythrocytes and found free in plasma.	Disease is usually sub-clinical. Clinical disease requires the presence of some other debilitating factors for manifestation. Fever and anaemia.	Neoarsphenamine @ 30mg/kg b.w and antimosan @ 6mg/kg. Imidocarb diproprionate.

Diagnosis

Parasitic eggs of trematodes, cestodes and nematodes as well as oocysts of *Eimeria* find their way out with the faeces of the host. So careful examination of the faeces will indicate the causative agent. However, for strongyle type of nematodes owing to their similar egg morphology, coproculture of faeces for harvesting the 3rd stage larvae from which identification can be made is necessary. Conducting post mortem of selected case for collection of adult parasites or their mucosal scrapings for larval stages shall also aid in the diagnosis of infection. Examination of infected skeletal and cardiac tissues after digestion by pepsin mixture shall rule out *Sarcocystis*infection. Examination of blood smears by different staining techniques shall detect blood protozoans like trypanosomes, *Babesia*, *Theileria*, etc. Application of Knotts technique will help in the detection of microfilariae of filaridnematodes in blood.Certain tests like Sabin-feldman dye test is used for detection of *Toxoplasma gondii* in blood. Collection of ectoparasites like ticks, lice, flies or their larvae from the body surface and preparation of their permanent mounts in the laboratory shall aide in the identification of arthropod infestations. However, for identification of mange mites, examination of skin scrapings is necessary.

Control

Regular but not frequent treatment of animals at proper dose level after assessing the parasitic load.However, the deworming and ectoparasiticidal schedule that has been framed as per the epidemiological data generated for a particular area should be adopted.

Prevent animals from grazing in marshy areas.

Silage and hay prepared from grasses cut from marshy areasshould be dried properly as metacercaria of flukes can survive in moist hay for up to 8 months.

Feeding troughs should be constructed in such a way that feed will not get contaminated by faeces.

Snail intermediate hosts can be controlled by application of molluscicides like coppersulphate, copper pentachlorophenate and N-tritylomorpholine and prevent animals from eating grasshoppers, ants, earthworms, beetlesthat act as intermediate hosts of trematodes, cestodes and some of the nematodes.

Infectivity on the pasture can be reduced by ploughing and reseeding the pasture or use of the pasture which has not been grazed the previous year.

Provide adequate drainage to marshy and low-lying areas like ponds, pools, ditches, etc. which will reduce population of biting flies like midges and mosquitoes.

Proper disposal of faecal debris along with strict sanitation of animal houses and yards.

Pasture resting.

Alternate grazing with hosts of different species (cattle/sheep/horse) should be encouraged.

Use of slurry is recommended.

Indoors of houses and animal sheds should be sprayed with longer residue activity insecticides to kill arthropod parasites or their larval stages.

Use of wire netting on doors and windows of animal houses should be encouraged to prevent insects from entering animal houses and use of electrocution grids in animal houses.

Spray of insecticides in the breeding places of insects like water logged area and marshy places to kill larvae and pupae of insects.

Spray of insecticides on the pasture to kill larvae and nymphs of ticks, which are found on blades of grass and in cracks and crevices of soil.

Repellents should be applied on the body of animals to keep insects away from them.

Cracks and crevices of animal sheds should be filled with tar or cement to reduce tick population, gamacid mites, bugs etc.

In short, the best practice to control parasitic infections in livestock is to follow integrated parasite/ pest management in letter and spirit.

Commonly used Anthelmintics

Name of drug	Dose	Types of worm	Route	Interval
Triclabendazole	10mg/kg b.wt.	F	Orally	Single dose (SD)
Rafoxanide	7.5 mg/kg b.wt.	F	Orally	SD
Oxyclozanide	15 mg/kg b. wt.	F, T		SD
Nitroxynil	10 mg/kg b. wt.	F	Subcutaneously	
Closantel	7.5-10mg/kg b.wt.	R, F	Orally	SD
Albendazole	7.5mg/kg b.wt.	R, F, T	Orally	SD
Fenbendazole	7.5mg/kg b.wt.	R, F, T	Orally	SD
Diamfenetide	120 mg/kg b. wt.	F	Orally	SD
Resorantel	65 mg/kg b. wt.	F, T		SD
Levamisole	7.5mg/kg b.wt. 10-20ml	R	Orally	SD
Levamisole + oxyclozanide	½ bolus/33.3kg b.wt.	R, F	Orally	SD
Mebendzole	10-15mg/kg b.wt.	R	Orally	For 3 days
Oxyclozanide + tetramisole	1ml/3 kg b.wt. or 2 boli/adult animal	F, R	Orally	SD
Ivermectin	0.2 mg/kg b. wt.	R	Orally or s/c according to the presentation	SD
Niclosamide	75-150 mg/kg b. wt.	T, F	Orally	SD
Praziquantel	15 mg/kg b. wt. or 1 tab/30kg b.wt.	T, F	Orally	SD

Note: Fluke=F, Round worm=R, Tapeworm=T

References

1. Soulsby EJL. *Helminths, arthropods and protozoa of domesticated animals.* 7. London: The English Language Book Society; 1982. pp. 126–173.
2. Thapar GS. Systematic survey of helminth parasites of domesticated animals in India. *Indian J Vet Sci.* 1956;26:211–271.
3. Yadav Anish, Khajuria JK, Raina AK. Seasonal prevalence of gastrointestinal parasites in sheep and goats of Jammu. *Indian Vet J.* 2006;20:65–68.

12 Commonly Occuring Plant Poisonings in Sheep and Goats in Kashmir

Syed Wasif Ahmad, Adil Mehraj Khan, Showkat Ahmad Shah

Symptoms of Plant Poisonings

- Sudden onset of obscure illness with no apparent cause.
- A number of animals showing acute disorders affecting gastrointestinal tract, central nervous system.
- Rapid loss of condition and prostration in case of chronic ingestion.
- Frequent observation of repeated attempts to void faeces, tympany, froth from mouth, respiratory distress and/ or coma.
- The lambs from deficient ewes may suffer from hypothermia and other deficiency syndromes viz. rickets, goitre, swayback, staggers, blindness. These effects are particularly marked in twin bearing ewes.
- Overfeeding in late pregnancy can produce disproportionately large single lambs and consequently a high incidence of dystocias particularly in maiden ewes.

Prevention and Control

- Remove promptly all animals from the infested area when plant poisonings strike.
- Treat sick animals under qualified veterinary supervision.
- Avoid early spring grazing as poisonous plants are early growers.
- Avoid turning very hungry animals to areas where toxic plants are expected to occur.
- First feed the animals on previously surveyed and safer areas.

Treatment

The poisoned animals are often detected in advanced stage of poisoning leaving little scope for any control measure. Specific antidotal measures are largely unavailable in plant poisonings except when cyanide and nitrate poisonings are indicated. Sodium nitrite (1g) and sodium thiosulphate (2g) in 15 ml water given subcutaneously constitute standard therapy of cyanide poisoning in sheep. Standard treatment in cases of nitrate/ nitrite poisoning is methylene blue (9 mg /kg body weight as 4 % solution i.v.) The treatment can be repeated in severe poisonings.

A saline purgative such as magnesium / sodium sulphate (60 mg/kg, orally) may be advisable if the animal is not suffering from severe gastroenteritis orfluid loss.

The convulsions may be treated with pentobarbitone (10-30 mg/kg,i.p) or diazepam (0.5-1.5 mg /kg , i.v.).Warm tea infusion acts as a cheap stimulant, and tea tannins precipitate alkaloids. Dilute potassium permanganate $KMNO_4$ (1:1000), Lugol's iodine or tincture iodine (5-10 drops /200 ml warm water) oxidize and detoxify alkaloids and other toxicants. Systemic detoxification and rehydration may be aided by parenteral dextrose–saline therapy; furosemide @ 1 mg /kg once or twice a day would facilitate detoxification.

Plant species	**Habitat & period**	**Toxic constituent**	**Observed effects**	**Remarks.**
Aconitum spp. *A.chasmanthum* (*mohri*) *A.deinorrhizum* (***Safedbish.Mohri***) 	Highland pastures June to September	Aconitine and related alkaloids	Tympany, froth from mouth, convulsions and death	All age groups are affected. young plants are more toxic, dry plants are also toxic.
Allium sp. (**Nai-ter**) 	Foothills March To May	Allyl and propyl disulphides	Acute diarrhoea, haemolytic anemia; haemoglobinuria; dyspnoea; death	Young sheep and goats are most affected
Colchicum luteum meadow saffron;autumn crocus (***irkimmoond***) 	Foothills April to May	Colchicine and colchiceine	Bloody dysentery and death	Young animals most affected; preflowering stage highly poisonous.

Dioscorea deltoids (***perath***) 	All pastures April to July	Dioscorine	Vomiting, convulsions haematuria and death	Fresh green leaves actually fatal to goats in particular.
Rhododendorncompanulatum (***yangipatra***) 	Alpine pasture June to September	Andromedotoxin (grayanotoxin)	Severe vomiting; dyspnoea; convulsions; sudden death	All age groups affected especially sheep.
Stipasibrica (***Ghomia, Ghumingassi***) 	Foothills and subalpine April to September	Hydrocyanic acid	Acute tympany; froth from mouth; convulsions; pyrexia death.	Only fresh green plants are toxic. Young animals are most affected.
*Sorghumhalepense*Johnson grass (***DrahamHassi-Gassi***) 	All places May to June	Cyanogenic glycoside and nitrate	Acute tympany; haematuria; dyspnoea; convulsions;blood bright red (cyanide) or chocolate brown (nitrite)	Hay safe fromcyanide but not fromnitrite; all age groups affected; goats eat roots and get poisoned.
Trifolliumrepense; white clover (***Batakloth;bataknuer or tri-patra***) 	All places April to May	Cyanide,goitrogen, nitrite oestrogens (isoflavones) and saponins	Acute tympany; bloat; reproductive disorders; possible death	All age groups affected preflowering green plants poisonous.

References

1. McCaughley W.J. Inorganic and Organic Poisons. In: Aitken I.D., editor. *Diseases of Sheep.* 4th ed. Blackwell Publishing; Oxford, UK: 2007. pp. 424–439.
2. Blood, D. C., Radostits, O. M., & Henderson, J. A. (1983). Veterinary Medicine, Bailliere Tindall.London.
3. Gracia, R., & Shepherd, G. (2004). Cyanide poisoning and its treatment. Pharmaco therapy: The Journal of Human Pharmacology and Drug Therapy, 24 (10), 1358-1365.

13 Vaccination in Sheep

Mir Nadeem Hassan, Iqra Hussain Shah, Ajaz Ahmad Ganie, Shaheen Farooq

Vaccines are an important part of disease prevention and control. If the risk of a particular disease is low, the insurance afforded by vaccination may not be required and the benefit provided might not be cost effective. If the risk of disease is high, the insurance afforded by vaccination may be very cost effective even if it doesnot completely prevent sickness.Most livestock vaccine and health management protocols revolve around the animal's stage of production. For sheep and goats, it is recommended to vaccinate prior to lambing, weaning, and breeding. The purpose of this publication is to offer a guide in establishing a health management schedule. Every operation is unique, and it is therefore imperative that producers consult with their veterinarian before establishing a specific vaccination and health protocol. The vaccination schedule for commonly reported diseases of sheep and goat of Kashmir region is depicted below.

Selecting the right vaccine

It is important to know which diseases animals should be vaccinated against. There are some common diseases that animals should be routinely protected against such. In some circumstances, disease should be diagnosed by a private veterinarian who will advise whether vaccination is recommended. Vaccine use should be part of a whole herd or flock health management program, incorporating biosecurity and disease prevention measures.

Some diseases may have several strains of causal organisms and some vaccines cover several disease complexes, so it is important to buy the correct vaccine for the situation. This is particularly relevant where more than one type of vaccine is available. It is also important to check that a vaccine is registered for the stock being vaccinated (e.g. there is a three-in-one and a six-in-one vaccine registered for sheep and lambs only, as well as two similar products which are registered for goats and kids in addition to sheep and lambs).

Some vaccines registered for the same disease but manufactured by different companies have different dosage regimes and booster recommendations.

Storing vaccines

As vaccines are biological products they will slowly lose their potency, even when stored under ideal conditions. The manufacturer has determined the expiry date for each batch of vaccine and provided it is stored correctly, it will retain its full potency up to the date shown.

Each vaccine has specific storage conditions. The majority of vaccines must be stored in a refrigerator where they are kept cold but not subject to freezing and thawing. Freezing may reduce the potency of some vaccines and may cause local reactions at the injection site.

Follow label instructions

The label on the vaccine carries important information about using the vaccine correctly: dosage; injection site; recommended vaccination program; and storage instructions. Users of veterinary chemicals are required by law to follow the label instructions unless they are using the chemical according to an approved permit or as directed in written instructions from a veterinarian. Vaccines are expensive and the label instructions should be followed to make the most of your investment.

Use sterile and safe equipment

Vaccines are usually administered by a syringe. Many vaccines are packaged in multi-dose containers for use with automatic syringes, which must be calibrated to deliver the right dose.

Needles should be sharp and as short as possible when vaccinating subcutaneously. Long needles may break or deposit the vaccine into the muscle instead of under the skin.

It is important that infection is not introduced when vaccinating animals. Equipment can be sterilised between use in a pressure cooker for 15 minutes, or by boiling in an open pan for 40 minutes. Chemicals, such as disinfectants or methylated spirits, should not be used as they may make the vaccine ineffective. After syringes and needles have been sterilised they should be kept in a covered container to protect them from dust.

Administering vaccines

In general, vaccines take 10-14 days to give protection. Vaccination should be part of a herd or flock health program. For vaccines to be most effective consider their use carefully in relation to the type of stock, season, previous property history and disease incidence e.g. young animals should be protected against the common and predictable diseases before management events such as castration, shearing, weaning and movement to new properties.

To get the most out of your vaccines protect them from heat or sunlight. Keep vaccines and equipment away from dirt and dust, which can contaminate equipment and introduce infection. Use a small table to help keep vaccines and equipment clean and off the ground.

Do not mix different vaccines together. Combined vaccines require a great deal of care in balancing the components. However, if more than one vaccine is required use separate syringes and administer them at different sites, at least 15 cm apart and preferably on different sides of the animal's body.

Avoid carcass damage by administering the vaccine according to label instructions and use the least commercially valuable site on the animal, for example high on the neck behind the ear.

When vaccinating a herd, ensure a full dose of vaccine is given to every animal. Check syringes carefully to ensure that the correct dose is being delivered. Give the required booster injection at the correct time to all animals that require it.

S. No.	Disease	Vaccination Schedule		Expected month for vaccination in a year	Trade name & Presentation Of companies	Dose & Route	Remarks
		Primary	Regular vaccination				
1.	Enterotox-aemia	At the age of 1 month if dam is vaccinated or at the age of 1 week if dam is unvac-cinated Booster dose is given after 15-21 days of primary vaccination	Pregnant ewes 1 month before lambing, Annually before rainy seasons	February	Multicomponent ET Vaccine or Raksha E.T. Vaccine (100 ml) Intervet/Indian Immunologicals	1 ml or 2 ml S/C	Regular vaccination along with booster dose is nec-essary
2.	Sheep Pox	At the age of 3 month & above	Annually	March or April	Raksha-SP Vaccine (50 ml) Indian Immunologicals Killed Sheep pox Vaccine	1 ml I/M 5ml s/c	Regular vaccination is necessary
3.	Goat Pox	At the age of 3 month & above	Annually	March or April	Goat Pox Vaccine IVRI	1 ml S/C	Regular vaccination is necessary
4.	PPR	4 month & above	Every 3 years interval	April	Raksha-PPR (100 doses) Indian Immunologicals	1 ml S/C	Regular vaccination is necessary
5	FMD	3 & 1/2 months& above	Biannually	March and September	FUTVAC	1 ml S/C or 1 ml I/M	Regular vaccination is necessary

Note: In chart, the primary vaccination schedule is maintained serially and for each vaccination a minimum of 15-21 days interval should be followed

References

1. Blood, D. C., Radostits, O. M., & Henderson, J. A. (1983). Veterinary Medicine, Bailliere Tindall.London.
2. Akan M., Oncel T., Sasreyyupoglu B., Haziroglu R., et al. Vaccination studies of lambs against experimental Mannheimia (Pasteurella) haemolytica infection. Small Ruminant Research. 2006;65:44–50.
3. Council Report Vaccination guidelines for small ruminants (sheep, goats, llamas, domestic deer, and wapiti) JAVMA. 1994; 205:1539–1544.
4. Lacasta D., Ferrer L.M., Ramos J.J., Gonzalez J.M., Ortin A., Fthenakis G.C. Vaccination schedules in small ruminant farms. 2015;181:34–48.

14 Abortions

Iqra Hussain Shah, Mir Nadeem Hassan, Asloob Ahmad Malik, Showkat Ahmad Shah, Ajaz Ahmad Ganie

Abortion is the early termination of pregnancy with expulsion of a dead foetus. A dead full term foetus is referred to as a stillbirth rather than an abortion. Abortions may occur days, weeks or months after exposure to an infectious microorganism, toxin, stress or genetic fault that impacts the foetus or placenta. This means abortions may be the result of factors occurring before animals entered the export process. Abortions in up to 3 % of breeders per year are considered normal.

Aborting animals may have higher risk of retained foetal membranes and uterine infection (metritis) if aborting occurs in late pregnancy, and development of the mammary gland may lead to mastitis.

Abortion is a significant problem in pregnant ewes/does and causes major financial losses for small holder livestock producers. It is a limiting production factor, as it decreases the potential number of replacement stocks for flock and milk production and increases the number of unproductive females maintained for long periods in the flock. Thus, abortion in sheep and goats significantly impacts the food security and livelihoods of rural smallholder farmers, as sheep and goats are an integral part of households by providing nutrition, employment and sources of income.

The following factors have been implicated, and should be considered if infectious causes are eliminated,

- Inadequate nutrition
- Pregnancy toxemia
- Stress
- Poor handling
- Vaccination
- Transport
- Dog worry
- Concurrent disease

- Pasteurellosis
- Chronic fluke

Clinical signs and diagnosis

Signs of abortion include straining, raised tail, red to brown coloured vaginal discharge, protrusion of foetal membranes, mammary gland development, and the discovery of a foetus or foetal membranes in a pen. Rectal and vaginal examination of animals may be required to confirm abortion.

Specimens for laboratory diagnosis ideally include acute and convalescent maternal sera and the complete fresh (or frozen) foetal-placental unit. Otherwise submit sections of foetal and placental tissues in formalin plus foetal stomach contents for microbiology.

In ewes 1-2%, but not beyond it are expected to abort each year due to non infections causes. But one thing must be remembered that any abortion must be assumed to be infections in origin and should be managed accordingly because only at the end of lambing one will be able to say that only 1% of ewes aborted and that all laboratory findings were negative.

Any infection during pregnancy causing toxemia, fever and malaise, even if for only a few days, can cause abortion. Thus infectious diseases like brucellosis, listeriosis, toxoplasmosis, vibriosis, blue tongue virus and sheep pox have potential to cause abortion in sheep.

Feeding of mouldy hay usually during winter months can lead to infection and production of toxins causing abortion.

Action is based on the knowledge that the aborted ewe is a source of infection for other ewes by means of her uterine discharges, placenta, fetus or lamb. Once abortion occurs one must assume that the aborted ewe is a danger for other ewes and the people. The thing becomes more important where lambing period extends over 2 months or more as in field conditions where an aborted ewe acts as a horizontal source of infection for other pregnant ewes, so;

1. Isolate and mark the aborted ewe.
2. Keep other ewes away from abortion site and disinfect the abortion site.
3. Seek Veterinary Assistance and send samples (lambs or fetus and placenta) to laboratory for examination.

If further ewes abort treat as above and send further samples to the laboratory. Do not assume that all abortions have same cause. Otherwise the aborted fetus as well as placenta should be buried deep under soil in such a manner that it is not dug out by stray dogs. Use hand gloves while handling the aborted material. Never handle aborted foetus with naked hands.

Upon confirmation of a particular cause, seek the advice of your concerned Veterinary Surgeon who may advise treatment of remaining pregnant ewes with a specific drug. This does not eliminate the infection but it can delay abortion, and thus increase the viability of lambs from infected ewes. Discuss the potential cost benefit ratio of this treatment with your veterinary surgeon before proceeding.

Keep away the pregnant women from handling a flock where an abortion has occurred.

For preventing abortion storm:

- Maintain a closed flock.
- Low level feeding of antibiotics.
- Prevent contamination of feed and water.
- Control cat population.
- Avoid stressful, overcrowded, and/or unhygienic conditions.

References

1. Dorsch MA, Cantón GJ, Driemeier D, Anderson ML, Moeller RB, Giannitti F 2021: Bacterial, protozoal and viral abortions in sheep and goats in South America: A review. Small Rum Res 205.
2. Quinn PJ, Markey BK, Leonard FC, Fitzpatrick ES, Fanning S, Hartigan PJ 2011: Veterinary Microbiology and Microbial Disease. Second edn, Wiley-Blackwell, United Kingdom, 928 p.
3. Sharma M, Batta K, Katoch RC, Andersen AA 2008: A field investigation of bacterial etiology of abortions among migratory sheep and goats in North-West hill states of India. VetArhiv 78: 65-71.

15 Management Year Calendar for Small Ruminants

Mir Nadeem Hassan, Manzoor Ahmad Yatoo, Andleeb Rafiq, Ajaz Ahmad Ganie, Showkat Ahmad Shah, Iqra Hussain Shah

January

- Root crops (Turnips/carrots) @ 500g/pregnant ewe.
- Pelleted feed @ 700gm/adult and 500gm/young stock.
- Supplementation of molasses @ 5-10% of concentrate mixture to prevent pregnancy toxaemia in pregnant ewes.
- Multicomponent Clostridial vaccination (MCC) to pregnant ewes before one month of expected date of lambing for protection against Lamb dysentry, Struck, Pulpy kidney disease, Black disease and Braxy. Provides passive immunity to lambs.
- Broad spectrum anthelmintic (pregnancy safe) dosing to pregnant ewes before 7-15 days of expected date of lambing.
- Routine recording of body weight during last three days of every month.
- Providing dry fodder (oats) in morning and evening @ 0.5 – 0.75 kg/animal from 2nd week.

February

- Segregation of pregnant ewes (18-20/pen).
- Preparation for lambing.
- Routine recording of body weight and morphometric measurements.
- Weak and under weight lambs along with dams are kept in warm pen (temperature 15-20 0C) for 1-2 days to reduce mortality from hypothermia.
- MCC vaccination to dry flock in last week.
- Creep mixture to lambs (15 days and above).
- Brand marking of male and female weaners.

March

- Amprolium dosing to lambs and kids (15-21 days age) @ 1gm/5kg body wt. Repeated after 21 days.
- MCC vaccination to lambs and kids (21-30 days age). Repeated after 15-21 days.
- Livestock left-out for grazing from 1st wk of March depending upon weather conditions and availability of pasture.
- Enhancement of Grazing period and gradual reduction in pelleted feed and dry fodder.

April

- FMD vaccination after 15-21 days of MCC.
- In lambs and kids, FMD vaccination after 15-21 days of 2nd dose of MCC booster.
- Broad spectrum anthelmintic dosing to flocks including lambs and kids (60-75 days age).
- Enhancement of grazing hours from 2nd wk with no supplementation.
- Shearing of livestock depending upon weather condition.
- Ectoparacidal dipping of livestock after shearing.

May

- Sheep pox vaccination to adult ovine flock and lambs, after 15-21 days of FMD vaccination. Repeated every year.
- Goat pox vaccination (from Mukteshwar, IVRI) to adult caprine stock and kids, after 15-21 days of FMD vaccination. Repeated every year.
- Weaning lambs attaining 90 days age (except weak lambs).

June

- PPR vaccination to lambs/kids after 15-21 days of sheep/goat pox vaccination. Repeated after 3 years interval.
- Weaning of lambs attaining 90 days age (except weak lambs).
- Migration to high land pasture, Ledarwas after 1st wk.
- Preparation of temporary tarpaulin sheet roofing with chain link fencing paddock at HLP to protect livestock from snow, rainfall and wild predators.
- Recording of monthly livestock body wt at HLP, Ledarwas.
- Cleaning and disinfection of paddock at HLP at regular intervals.

July

- 2nd dose of MCC vaccination to ewes (after six months of 1st vaccination).
- Cleaning and disinfection of sheep sheds at MRCSG Campus.

August

- 2nd dose of MCC vaccination to ram, weaners and caprine stock (after six month of 1st vaccination).
- Preparation of culling list.
- Preparation of breeding plan.

September

- Downward migration after 1st week.
- Shearing (autumn clip) from 2nd week.
- Ectoparacidal dipping after shearing.
- Proper anthelmintic dosing before breeding.
- Inspection & marking of ram and ewes (maiden ewes body weight > 28 Kg) before breeding.

October

- Marking of breeding rams daily.
- Daily recording of tupping (mark on the rump of ewes) up to three cycles/51 days.
- Auction/sale of livestock to development departments, NGO's and progressive farmers. Concentrate feeding to weaners.

November

- Providing bhusa after second week for meeting the requirements(insufficient biomass available for surface grazing).
- Providing pelleted feed initially @ 200 gm/adult and 150 gm/young from second week. Increased gradually.

December

- Extensive grazing period decreased gradually.
- Providing sorghum @ 0.5 kg/animal after depletion of bhusa and pelleted feed @ 300 gm/adult and 200 gm/weaner from last week.

- Intensive management (Stall feeding) from last week.
- Sanitation and cleanness in and around the livestock sheds before stall feeding.
- Increasing dry fodder (sorghum) up to 1 kg/animal and pelleted feed up to @ 500 gm/adult and 400 gm/young stock.

References

1. Goat Production and Health Management" by S K Jindal
2. Hand book of Animal Husbandry. V.K. Taneja, C. Chakravarty, C.S. Viswanath, Aruna T. Kumar. Third revised. Indian Council of Agricultural Research, New Delhi.

16 Methods of Drug Administration in Sheep

Syed Ashiq Hussain

The administration of drugs is a common farm health procedure. The correct technique of drug administration enhances the farm economy and sometimes save the life of an animal but an incorrect technique could result in aspiration pneumonia and injection site lesions. Always use proper procedures, equipments and dosage of drugs for effective medication. The various methods of drug administration are drenching, bolus/capsule and parenteral injections.

1. Calculation of Drug Dosage

It is always best to use correct dosage of each drug for any route of administration. The incorrect dosage can result into underdosing or overdosing. Dosage should be calculated according to thc dose rate of drug or the manufacturer's recommendations based on the weight of the animal. The following example shows the steps of dosage calculation for administering oxytetracycline in sheep.

- Weight of animal= 40 kg.
- Dose of oxytetracycline=10 mg/kg bodyweight.
- Dosage = Weight of animal X dose rate of drug
- In our example, dosage = 40kg x 10 mg/kg = 400 mg.

The next step is calculation for the volume of drug (in ml) required by the animal. this step is for liquid form of drugs only. If the label of Oxytetracycline states that the injectable solution contains 50 mg/ml. It means that 50 mg of actual medicine is available in each ml of the solution.

- Volume of the drug required = dosage divided by strength of the medicine. In present example, volume = 400 mg/ 50 mg per ml= 8 ml of Oxytetracycline

Tips for Dose Calculation

- 1% solution of any drug means that 10 mg of that drug is present in 1ml of injectable solution. Similarly, 10% concentration means 100mg/ml, and so on.

2. Drenching

- Small volumes of liquid (<30 ml) are frequently administered by mouth using drench gun or an oral syringe with a long metal nozzle for drenching. Now a days automatic drench guns are also available.
- The animal should be suitably restrained with the hindquarters lodged so that sheep cannot move backwards from the operator.
- Stand behind the sheep (if you do not have a chute or an assistant to hold the sheep, you may wish to back the sheep into a corner to prevent it from backing up as the dose is given).
- Place a hand under the animal's head and tilt it slightly to the side. The gun nozzle should be placed in the gap between the molar and incisor teeth and then the liquid is squirted over the back of the tongue. The animal should swallow before the gun is removed. Covering the animal's nostrils will encourage swallowing.
- Do not insert the nozzle toofar into the throat. The trachea (windpipe) lays directly beneath the oesophagus (leading to the digestive system). If the liquid is forced into the trachea, the sheep will inhale the drench.
- Larger volumes of medicines can be administered by stomach tube.
- Precaution during drenching: Poor drenching technique can cause aspiration pneumonia (inflammation of the lungs caused by breathing in foreign material). Aspiration can be prevented by
    - Holding animal's head horizontally and it should not be tilted up.
    - Drench guns should not be forced down the animal's throat.
    - Drenches should not be administered faster than the animal can swallow.

3. Administering Bolus/Capsule

- This means the administration of individual doses of medication in pill form.
- Ideally inserted by the bolus gun (or balling gun). A balling gun is a plastic or metal tube with a capsule holder on one end and a plunger on the other end (to eject the capsule).
- Under field conditions, the bolus or capsule can be also given manually. The bolus should be deposited over the base of the tongue. Otherwise, the pill may be chewed or spit out by the animal.
- Ensure proper size boluses are used in sheep as big cattle pills could become stuck halfway down the oesophagus and cause choking.

4. Parenteral Administration of medicines

Medicines can be given by intravenous, intramuscular, subcutaneous and intraperitoneal injection. A new sterile needle and syringe should be used on every occasion. The only exception to single syringe/needle use is the administration of a vaccine to a number of sheep using an automatic syringe. The correct size of hypodermic needle should be used as mentioned below

Lambs < 10 Kg	21guage
Lambs 10-20 Kg	20 guage
Lambs 20-40 Kg	19 guage
Adult sheep	18 guage

Thevarious methods of parenteral drug administration are

4.1. Intravenous Injection

- For intravenous (IV) injections, the medication is administered into a vein, so get advice and training from your veterinarian.
- IV injection requires skill to locate the vein.
- Ensure that the entire drug enters the vein; the drug spillage around the vein can cause its inflammation.
- Remember that timely IV injection may be the only means sometimes in emergency situation or certain diseases to save an animal. But if not done carefully but too quickly it could cause death.

4.2. Intramuscular (IM) Injection

- This is the most common method of parenteral drug administration.
- The injection is given deep into a muscle at the recommended sites.
- Usually 1 inch long needle is enough to penetrate the skin, subcutaneous tissue and fat to reach the muscle.
- The needle should enter at 90 degree angle to the body.
- Before administering the drug, ensure you have not hit a vein. In that case blood will be withdrawn into the syringe and you need to re-try a different location.
- The drugs administered by intramuscular route absorb faster than subcutaneous but slower than intravenous route.

4.3. Subcutaneous Injection

- Use a 0.5 to 1 inch long needle.

- The subcutaneous (SC) injections are given into a tent of skin. Lift a fold of skin and insert the needle through the skin into the tented space.
- Ideally the needle enters the skin at an angle of 30 to 45 degrees to the body.
- Occasionally there may be lump formation at the injection site, this not a generally problem and will disperse. This occurs particularly with large doses.
- Choose SC when given a choice of IM or SC on the product label.

4.4. Intraperitoneal Injection

- For intraperitoneal injection also get advice and training from your veterinarian.
- In severe hypothermia of lambs or kids intraperitoneal or intravenous routes are used to save the animal.

Caution and Precautions

- When giving injections always get veterinary advice to make sure the products are appropriate and you know the correct procedure.
- Various pathogenic bacteria are present on the surface of the skin which may cause infection if injected with the medication. Therefore; when time allows or for valuable animals, or if the environment is very dirty, take every care to clean and disinfect the skin before injecting.
- If the injection is made on the side of the neck and the site is covered by wool, close clipping of the skin is ideal, but careful separation of the fleece and scrubbing the skin with a disinfectant like tincture of iodine may be possible. A fresh swab of disinfectant should be used for each animal.
- If you have to vaccinate good number of animals in one go you may not have time to disinfect the skin of each sheep. You may simply proceed down the race injecting each sheep in a clean part of the neck, taking care to keep your hands clean. Also take care to keep the needle clean and replacing it if it becomes dirty.
- As the rapid injection of any medication into a vein can be lethal, all intravenous injections are given very slowly while the sheep must also be well restrained.

4.5. Sites for Injections

- Intravenous injections are given into jugular vein.
- Intramuscular injections are commonly given in the triangular area of the neck, in front of the shoulder. Where possible IM injection should be given in the neck rather than in hind quarter as abscess in the later area may result in loss of prized meat cut if the animal is later slaughtered.

- The preferred site for SC injections is skin just behind the elbow (or the loose skin of the axilla (armpit). They can also be given in the triangular area in front of the shoulders between the top and bottom of the shoulder blade and corner of the jaw.

4.6. Points to remember for parenteral drug administration

1. Wash your hands before and after handling the products.
2. Use sterile/disposable needles and syringes.
3. Use only hot water to rinse syringes before using modified live virus vaccines. Chemicals may destroy the live virus.
4. Choose the correct length needle.
5. Shake the medicine bottle and swab its cap with clean surgical spirit or 70% alcohol.
6. For drawing liquid from a new bottle into a syringe, it is often hard to suck out because there is no air in the new bottle or container. Draw slightly more amount of air into the syringe than that will be replaced by the injection, and inject this air into the bottle through the rubber cap. This will put enough air into the bottle and allows you to fill the syringe. You may have to do this every now and again if you find it hard to fill the syringe. Or you can put another needle in the cap of the container to act as an air entry while you are drawing product out.
7. Draw slightly more liquid into the syringe than required.
8. Push the plunger slightly to expel any air bubbles and adjust to the correct dose.
9. Detach the syringe, leaving the needle in the cap for next dose withdrawal.
10. Attach a second needle to the syringe and expel any air from the needle.
11. Split large volumes into smaller amounts and inject in different locations (e.g. opposite side of the neck).
12. For IM injections, inject no more than 10 m1 per site. For SC injections, inject only 20 ml per site.

References

1. Blood, D. C., Radostits, O. M., & Henderson, J. A. (1983). Veterinary Medicine, Bailliere Tindall.London.
2. Radostits O.M., Gay C.C., Hinchliff K.W., Constable P.D. Veterinary Medicine: A Textbook of the Diseases of Cattle, Horses, Sheep, Pigs and Goats. tenth ed. WB Saunders; NY: 2007.

17 Pasture Management

Dr Manzoor Ahmad Yatoo, Ajaz Ahmad Ganie, Mir Nadeem Hassan, Iqra Hussain Shah

Pasture management is key to grazing because livestock productivity during the season depends on pasture efficiency. It depends on many factors: quality and quantity of forage, grazing, natural conditions, etc. Competent pasture management systems help to optimize the complex process.

What Is Pasture Management?

Pasture management is a practice for profitable livestock production providing the animals with forage grasses and legumes and keeping the soil healthy. It defines strategies to improve grass health and forage production, maintain a healthy ecosystem, and reduce production costs. Success depends on knowing how all the elements are interconnected. For example, cattle grazing in pasture affects the growth of grass, which in turn affects the number of weeds. Water flow for farm production affects the environment and animal health. Thus, pasture improvement depends on the effective control of each aspect.

Why Is Pasture Management Important?

The importance of pasture management is that it improves the health and sustainability of the ecosystem. At the same time, a poorly organized system causes weed invasions, slows the recovery of forage grass, and reduces quality. Pasture renewal may solve these problems temporarily, but for the long-term development of the farm, it is much more effective to optimize grassland management practices.

Well-managed grasslands are the key to maintaining long-term livestock productivity.

How To Benefit From Pasture Management?

A sound strategy ensures better farm profitability, ecosystem, and animal health. Here are just several advantages of well-organized pasture maintenance:

- reduced weeds;
- enhanced soil drainage;
- improved water quality;
- good distribution of nutrients (manure);
- decreasing the use of supplementary nutrition (silage and hay).

Overall, pasture care and maintenance are very beneficial economically. This practice reduces labor costs, strengthens the soil tilth, reduces erosion risk, and gives an economical feed source. Moreover, grazing provides the physical exercise necessary for the animals' health.

At the same time, pasture management should be a permanent process. It is the only way farmers can maintain the area's optimal condition. For example, there is always the risk of overdevelopment of forage grass. Consequently, grazings accumulate too much water, which reduces the amount of fiber and deteriorates the quality of livestock nutrition. Grassland maintenance implies regular monitoring and timely eradication of noxious plants, as well as forecasting weather changes to avoid negative consequences. The problems of pasture management contain the complexity of organizing the process: it requires careful planning, usually a year in advance. However, such a system's advantages are indeed more significant.

- Migrate in stages and do not cover long distances in shorter time
- Approach Chopan/ Chopan association through breeder association to negotiate terms, especially the issue of missing animals on return from pasture so that proper accounting occurs.
- Ensure all your animals have proper brand numbers and properly recorded photographs for better accounting and dealing with the problem of missing animals.
- It is usually a great adventure and to visit highland pastures. Poets, mountaineers, general tourists, children and other adventurous and nature loving persons have been treasuring visits to high lands or pastures. Even if labour has been engaged for rearing one's livestock at HLP or Chopans manage them there a breeder/entrepreneur should sometimes visit his livestock.
- Silvipasture/Agro-forestry

Management of pastures and Silvipasture needs community awareness and organized effort by breeders. Breeder societies and cooperative structures are needed for maintenance and development of pastures and preventing further encroachment. The department through its forage production wing is encouraging farmers to cultivate good quality grasses/legumes and perennial grasses for their livestock and seeds of such improved varieties of grasses are being distributed amongst the farmers for fodder development. Further establishment of silvi pastures for increasing bio-mass by introducing quality grasses and fodder trees is also encouraged.

Pasture of migratory route should be developed by involving the nomadic grazers through providing allied facilities of drinking water for livestock, improvement of bridle paths and constructing of a few shelters sheds on the migratory route.

Breeders are advised to contact agriculture wing of the Department of Sheep Husbandry, Lalmandi for planning fodder production and various schemes for better fodder production including provision of quality seeds for fodder.

Breeders are advised to have good familiarity with the grasses locally available and quality of pastures in which their flocks graze. They should get proximate analysis of feed and fodder done in the laboratory.

References

1. Conant RT, Paustian K, Elliott ET. Grassland management and conversion into grassland: Effects on soil carbon. Ecological Applications. 2001;11(2):343-355.
2. FAO. Challenges and opportunities for carbon sequestration in grassland systems—A technical report on grassland management and climate change mitigation. Integrated Crop Management. 2010.
3. King WM, Rennie GM, Devantier B, Hoogendoorn CJ. Impacts of grazing management on hill country pastures: Principles and practices. Hill Country – Grassland Research and Practice. 2016;16:203-212

18 Commonly Used Drugs in Sheep and Goats

Iqra Hussain Shah, Mir Nadeem Hassan, Syed Wasif, Syed Ashiq Hussain, Showkat Ahmad Shah

Drugs Acting on Alimentary Tract

Trade name	Name of drug	Dose	Route	Interval
Neblon	Indigenous	6-10g	Orally	BID
Cflox –Tz	Ciprofloxacin + tinidazole	1tab/25kg b.wt	-do-	O.D
Svifta	Phthalyl sulphathiazole + pectin + koalin	2-4 boli	-do-	BID
Belamyl	Thiamine hydrochloride + riboflavin + liver extract	0.25-0.5ml	I/M	O.D
Brotone	-do-	20 ml	Orally	O.D
Anorexon forte	Copper sulphate + ferrous sulphate +	½ bolus	Orally	O.D
Bovirum	Antimony potassium tartarate + $CuSO_4$	1-2 boli	Orally	O.D
Onfeed	$CuSO_4$ + nicotamide	1 boli	Orally	O.D
Rumisaac	Amylase + cellulose + lactobacillus	2 boli	Orally	O.D
Hb strong	Indigenous	2.5g	Orally	BID
Himalayan batisa	-do-	15g	Orally	-do-
Rumbion bolus	-do-	1 bolus	Orally	-do-
Sulfa bolus	Sulphadimidine	1 bolus	Orally	-do-
Inj. Sulphadimidine (333mg/ml)	-do-	10ml	I/V	-do-

Antidotes and Haemostatics

Trade name	Name of drug	Dose	Route	Interval
Inj. Atropine sulphate (1mg/ml)	Atropine sulphate	0.5mg/kg	I/V, I/M	4 hourly
Styplon	Indigenous	½ bolus	Orally	BID
Revici	N-butanol citric acid	5-10 ml	I/V ,I/M	SOS

NSAIDS

Trade name	Name of drug	Dose	Route	Interval
Melonex Melonex (power) 20mg/ml	Meloxicam	0.5mg/kg (pneumonia) 0.2-0.3 mg/kg	I/M, I/V, S/C, oral	O.D
Zobid-M (5mg/ml)	Meloxicam	-do-	-do-	O.D
Proxyvet –MP	Meloxicam+ paracetamol	30ml/300kg.b.wt	-do-	O.D

Bronchodilators

Trade name	Name of drug	Dose	Route	Interval
Inj. Deriphylline: 50.6mg/2ml	Theophylline	6-11mg/kg b.wt	I/M, I/V	TID
Caflon	Indigenous	6-12g	Orally	BID, TID

Drugs Acting on Urogenital System

Trade name	Name of drug	Dose	Route	Interval
Inj. Lasix: (20mg/ml, 10mg/ml)	Frusemide	1-2mg/kg b.wt	I/V, I/M	SOS
Inj. Pitocin (5IU/0.5ml)	Oxytocin	30-50 IU	I/M, I/V	SOS
Inj. Epidosin: (8mg/ ml)	Valethamate:	10-20mg	I/M, I/V	SOS
Replanta	Indigenous	16-24g followed by 8-12g	Orally	QID
Cystone	Indigenous	2 tabs	Orally	TID
Furex	Nitrofurazone + urea	1-2 boli	I/U	SOS
Furex-M	Nitrofurazone + urea+ metronidazole	2 boli	I/U	SOS

Cleanex	Nitrofurazone + urea+ metronidazole +povidine iodine	2-4 boli	I/U	SOS
Lixen IU 1.5g	Cephalexin	Reconstitute with 60 ml sterile water and infuse to uterus	I/U	SOS

Anthelmintics

Trade name	Name of drug	Dose	Route	Interval
Albendos (3g bolus 2.5%soln.)	Albendazole	5-10mg/kg b.wt	Orally	SD
Panacur (Bolus 1.5g Powder 250mg/g)	Fenbendazole	5-7.5mg/kg b.wt	Orally	SD
Wormal 30% w/w	Levamisole	7.5mg/kg b.wt 10-20ml	Orally	SD
Fasmin	levamisole and oxyclozanide	½ bolus/33.3kg b.wt	Orally	SD
Wormin 500mg/5g	Mebendzole	10-15mg/kg b.wt	Orally	For 3 days
Nilzan (3g bolus)	Oxyclozanide + tetramisole	1ml/3 kg b.wt or 2 boli/adult animal	Orally	SD
Prazfen	Praziquantel + fenbendazole	1 tab/30kg b.wt	Orally	SD
Rafoxin (2g bolus)	Rafoxanide	7.5 mg/kg b.wt	Orally	SD
Azolanide (Susp 90ml)	Rafoxanide +Albendazole	1ml/3 kg b.wt	Orally	SD
Bolus Fasinex (900mg)	Triclabendazole	10mg/kg b.wt	Orally	SD
Flukeout	Triclabendazole + ivermectin	1 ml/5kg b.wt	Orally	SD
Zycloz (1000 mg, susp 15% w/v)	Closantel	7.5-10mg/kg b.wt	Orally	SD

Antiprotozoals

Trade name	Name of drug	Dose	Route	Interval
	Metronidazole			
Centrogyl LM	Metronidazole + furazolidone + loperamide Hcl	1-2 boli/day	Orally	O.D
C-flox-TZ	Tinidazole + ciprofloxacin	1 tab/15-25 kg b.wt	Orally	BID
Powerflox TZ	Norfloxacin + tinidazole	½ -1 bolus	Orally	BID
Butalex Inj. 50mg/ml	Bupravaquone	2.5mg/kg b.wt	I/M	Stat
Berenil RTU	Diminazine aceturate	1ml/10-20 kg B.wt	I/M	Stat
Anthiomaline Inj. 6% w/v	Lithium antimony thiomalate	15-20 ml	I/M	3 doses in weekly interval

Anticoccidials

Trade Name	Compound	Indications	Dose
Amprolium aquinoxiline powder	Amprolium +Sulfaquinoxiline	Coccidiosis	2 gms./ litre of water
Amprolium powder	Amprolium Hcl 20 %	Coccidiosis	Lambs 1 gm/2Ls of Water
Coxicure	Sulfaquinoxillin 42.5 % + Diaverdine 7.5%	Coccidiosis	Lambs 2gm/kg feed

Antifungals

Grisovinfp	Griseofulvin 125 mg. /TAB	Antifungal	2-3 tabs per day

Antibacterials

Trade name	Name of drug	Dose	Route	Interval
Intamox	Amoxicillin + cloxacillin	10mg/kg b.wt	I/M, I/V	BID
Amoxirum	Amoxicillin + sulbactam	300 mg	I/M, I/V	O.D, BID
AC-Vet	Ampicillin + Cloxacillin	2 g	I/V, I/M	QID

Longacillin – Vet 24	Benzathine penicillin	24 lac	I/M	
Intacef	Ceftriaxone	15-25 mg/kg b.wt	I/M, I/V	BID
Cifran Inf 100mg/50ml	Ciprofloxacin	4-5 mg/kg b.wt	I/M, I/V	
Meriquin Inj. 100mg/ml Floxidin Inj. 10%w/v	Enrofloxacin	2.5-5 mg/kg b.wt	I/M,	BID
Fortified procaine penicillin 40 Lac	Fortified procaine penicillin	4000-10000 IU/ kg b.wt	I/M,	O.D
Oxytetracycline 50mg/ml	OXY-50	5-10mg/kg b.wt	I/M, I/V	O.D
Streptomycin + penicillin	Dicrysticin –S	1ml/5kg.bwt	I/M, I/V	BID
Sulphadiazine + trimethoprim	Biosulpha	15-30mg/kg b.wt	Oral, I/M, I/V	
Sulphadimidine 5g bolus, inj. 333mg/ml	Sulphadimidine	100 mg/kg b.wt	Oral, I/V	BID
Sulfagut Bolus 5g	Sulfaguanidine	100-300 mg/kg b.wt	Oral	BID
Stylosin -100 premix	Tylosin	10-20 mg/kg b.wt	Oral	BID

Ectoparasiticides

Trade name	Name of drug	Dose	Route	Interval
Taktic 12.5%	Amitraz	2ml/lt of water for ticks 4ml/lt of water for mites	Topical application	SOS
Clinar	Cypermethrin	1ml/L of water (body spray) 5 ml/L of water (back line spray)	Topical application	SOS
Butox 12.5mg/ml	Deltamethrin	2-3ml/lt of water for ticks 4-6ml/lt of water for mites 1-1.5ml/lt of water for lice	Topical application	SOS
Lorexane	Gamma benzene hexa chloride	Spray on affected area	Topical application	SOS
Inj. Ivermectin Inj. Hitek	Ivermectin	0.2mg/kg b.wt	Sub cut	7 days

Nutrition and Metabolism

Trade name	Name of drug	Dose	Route	Interval
Veta A injection (6 lac IU/2ml)	Vitamin A	3-5 ampules	I/M	Weekly
Agrimin forte	Vitamin A, D&E, copper, nicotamide, cobalt, zinc, magnesium, iron, selenium, manganese, phosphorus, sulphur	5-10g	Orally	O.D
Inj. CALDEE-12	Calcium levulinate, cholecalciferol & vit. B	1-3 ml	I/M	Weekly
OSTOPHOS-FS	Calcium, phosphorus and microcrystalline hydroxy apatite complex	10g	Orally	O.D
Liq.Vimeral	Vit. A,B_{12},E &D	5ml	Orally	O.D
Inj. Tonophosphan	Phosphonic acid	1-3ml	I/M, I/V	O.D
Inj. Tribivet	Vit $B_{1, 6, 12}$	0.5 ml	I/M, I/V	A.D
Inj. Imferon	Iron dextran 50mg/ml	100-200mg	I/M, I/V	Weekly, 2-3 inj.
Inj. Jecktofer	Iron-sorbitol-citric acid complex eq. to 50mg	100-200mg	I/M, I/V	Weekly, 2-3 inj

Antihistamines and Antialergics

Avil vet inj.	Pheniramine maleate 22.75 mg/ml	Purities, Dermatitis, Rhinitis, Edema, Insect Bite, Anaphylaxis	0.5-2ml I/M or I/V
Betacortil Inj	Beta methasone Sod.4mg/ml	Ketosis, Arthritis, Stress, Allergy, Shock	1 ml I/M Contraindicated in Advance pregnancy
Curadox inj.	Dexamethasone Sod. Phos.4mg/ml	Shock, Lameness, Arthritis, Stress, Ketosis, Allergy	1 ml I.M Contraindicated in Advance pregnancy
Prednisolane	Predisolone 1%	Anti-inflammatory, Stress, Arthritis, Allergy, Mastitis	2.5-5 ml IM

Drugs for Metabolic Disorders

Calgonate inj	Cal. boro gluc. 25%	Hypocalcaemia' lactation tetani liver damage,milk fever.	25-50ml I.V slow
Calcium boro gluconate Inj.	Cal.boro gluconate 25%	Hypocalcaemia, lactation tetani, milk fever	25-50 ml I.V. slow
Inj.Dextrose 5%	Dextrose 5%	Pregnancy toxaemia, Hypoglycaemia	50 ml. I.V.
Inj. D.N. S.	Dextrose 5%, NaClo.9%	Dehydration hypoglycemia, surgery	Dose as per requirement I/V
Inj Normal Saline	Sod. Chloride0.9%	Dehydration, Shock, fluid loss	Dose as per requirement I/V
Rintose Inj.	Nacl 6 gm+Kcl 0.4gm+Cal chloride 270 mg+Sod.Lactate 3.12gm+ Glucose 200 gm/Lit	Ketosis, Dehydration, ,Diarrhoea	100-200 ml/Day for 2-3 Days as per requirement

Herbal Drugs

DRUG	INDICATION	DOSE
Caflon powder	Expectorant, cough	6-12 gm with jiggery
Cystone powder and tablets	Urinary calculi, Hemeturia,	1-2 gms TID PO 2 tabs TID PO
Diarex tabs	Antidiarrhoeal	2-4 tab TID
Galog	Galactagogue,Agalactia	10 gm PO daily
Herminsa powder	Stomachic, Anorexia, Dyspepsia, Indigestion, Constipation,	20 gm daily with feed
Himalyan Batisa	Digestive appetizer, tonic, constipation, flatulence,	10-15 gm daily with feed
liv-52 powder,tabs, liquid	Liver tonic anorexia, jaundice	3-6 gm daily PO 2-3 tabs daily PO
Neblon	Antidiarrhoeal, scours, dysentery	6-10 gm orally
Tentex forte	Impotance, decreased libido	1-2 gm daily for 7 days for rams and bucks
Timpol	Bloat	20-25 gm PO BID with 250 ml warm water or 250 -500ml Linseed oil 3-4 hrly

References

1. Blood, D. C., Radostits, O. M., & Henderson, J. A. (1983). Veterinary Medicine, Bailliere Tindall.London.
2. Smith M.C., Sherman D.M. *Goat Medicine.* second ed. Wiley-Blackwell; Ames, IA: 2009.
3. Radostits O.M., Gay C.C., Hinchliff K.W., Constable P.D. *Veterinary Medicine: A Textbook of the Diseases of Cattle, Horses, Sheep, Pigs and Goats.* tenth ed. WB Saunders; NY: 2007. ISBN 978-0702-07772.